6
7
8
9
16
17
18
19
20
26
27
28
29
30
36
37
38
39
40
46
47
48
49
50
56
57
58
59
60
MW00558753

The Science of Cleaning

90 21 17

The Science of Cleaning

Use the Power of Chemistry to Clean Smarter, Easier, and Safer

* WITH SOLUTIONS FOR EVERY KIND OF DIRT *

DARIO BRESSANINI, PhD

Translated by Denise Muir, Victoria Weavil, Ailsa Wood, and Marinella Mezzanotte

Illustrated by Beth Bugler

NEW YORK

Originally published in Italy as *La Scienza delle Pulizie* by Feltrinelli Editore in 2022. First published in English in North America by The Experiment, LLC, in 2024.

The Experiment, LLC | 220 East 23rd Street, Suite 600 | New York, NY 10010-4658 | theexperimentpublishing.com

Library of Congress Cataloging-in-Publication Data

Names: Bressanini, Dario, author. | Muir, Denise, translator. | Weavil, Victoria, 1984- translator. | Wood, Ailsa, translator. | Mezzanotte, Marinella, translator.
Title: The science of cleaning : use the power of chemistry to clean smarter, easier, and safer-with solutions for every kind of dirt / Dario Bressanini, PhD ; translated by Denise Muir, Victoria Weavil, Ailsa Wood, and Marinella Mezzanotte.
Other titles: Scienza delle pulizie. English
Description: New York, NY : The Experiment, 2024. | Includes index. | Translated from the Italian.
Identifiers: LCCN 2023055146 (print) | LCCN 2023055147 (ebook) | ISBN 9781891011320 | ISBN 9781891011337 (ebook)
Subjects: LCSH: Cleaning compounds. | House cleaning.
Classification: LCC TP990 .B8513 2024 (print) | LCC TP990 (ebook) | DDC 668/.1--dc23/eng/20240111
LC record available at https://lccn.loc.gov/2023055146
LC ebook record available at https://lccn.loc.gov/2023055147

ISBN 978-1-891011-32-0
Ebook ISBN 978-1-891011-33-7

Cover and text design, and illustrations throughout, by Beth Bugler
Author photo by Barbara Torresan
Translation by Denise Muir, Victoria Weavil, Ailsa Wood, and Marinella Mezzanotte

Manufactured in Turkey

First printing April 2024
10 9 8 7 6 5 4 3 2 1

+3NaOH/H2O
3R
ONa

CONTENTS

INTRODUCTION

Maybe it's just me, but when I shop at the supermarket, the most stressful aisles to navigate are without a doubt the household cleaning and hygiene ones. There are row upon row, mile upon mile of cleaning products whose packages look like they're singing and dancing for my attention.

It really shouldn't be that difficult: My sweater's got a stain on it and I want something to remove it. But what? Which of the hundreds of options should I go for? Standing there in the cleaning aisle, it feels like the multitude of bottles and boxes and containers are all staring at me, whispering, "Buy me, buy me." Liquid or powder? Hand-wash or machine? "With enzymes" or "without" or "removes stains in cold water"? Does it matter? And what about the water: hot, warm, or cold? What's the difference between products for cotton, wool, polyester, or some other material? Help!

It's the same story in the next aisle: even more rows stacked high with all kinds of dishwashing liquids. It's not easy to know which to buy, and you very often go home wondering if you got the wrong one. This feeling of cluelessness occurs no matter what part of your home you want to clean: floors, windows, curtains, bathroom, sink, stovetop, oven, pots, or pans.

A book about cleaning would have been much shorter seventy years ago, when there

weren't so many specialist products to worry about. In less than a century, though, separate cleaning solutions have appeared for washing dishes, scrubbing floors, laundering clothes, and even keeping ourselves clean. Aside from when people used to wash with ash (more about that later), a trusty old bar of soap was once the answer to everything. The only exception might've been special soap flakes for washing clothes or cleaning floors, but at the end of the day, everything was cleaned with soap, often the same bar.

The invention of modern detergents and their widespread commercialization, combined with the increasing use of home appliances like washing machines and dishwashers, has fueled the expanding array of cleaning products. While such an eclectic assortment certainly makes life much easier for those seeking to keep their homes clean, it also makes decisions a lot more complicated. Quite often, you end up with the annoying feeling that a lot of the products do the same thing but in different packaging, and that manufacturers just want us to buy multiple versions to take advantage of how little we know about the chemicals they're made of.

At one stage of my life, I'd just grab the first thing that caught my eye. I couldn't tell the many types of soap apart, and despite being a professional chemist, I didn't bother looking at the chemical composition. Not anymore. I've since learned that cleaning products' overwhelming packaging hasn't changed the chemical principles underpinning them—they're still the same as they were a hundred years ago. That's why I decided to write this book to help you understand what you need to clean the things you want to clean. I've learned that I don't always need a specific product, as sometimes I can use what I already have on hand—a homemade concoction that doesn't have to cost a lot of money will often do the trick. (And let's be honest: Cleaning products aren't cheap.) But ultimately, there will also be times when something specific, devised to fulfill a particular purpose by chemists who are more qualified than me, is just the thing for the job.

I have also noticed two seemingly contradictory trends in consumers. On the one hand, people want out-of-the-box solutions for specific problems: for example, a specialist product that removes a particular kind of stain from a particular kind of fabric in a preferred way (whether that's machine- or hand-wash). After all, life is complicated enough without having to work out how to get rid of a grease mark, clean the toilet, or unblock the sink. Can't we just buy the right product for the job? On the other hand, this contrasts with the quest for a one-solution-fits-all cleaning product: something that works just as well for the stain, the toilet, and the sink. Oh, and without polluting the environment, harming kids or pets, or denting the household budget—while being one hundred percent effective. And if I can make it at home from a mixture of salt, vinegar, baking soda, and lemon, even better.

Unfortunately, fulfilling both wishes would be just about impossible—a specific product for every potential cleaning requirement is never going to happen. The number of potential combinations would fill not just tens, but hundreds of supermarkets. The flip side is that we'll never have just one product that cleans everything at once, either. There's a chemical reason for that, and I'd like to share it with you right here in this book. Thousands of chemists have spent decades devising the most effective cleaning solutions. As I will explain, there's room for salt, vinegar, baking soda, and lemon in our housekeeping solutions, but the idea that they could replace the whole spectrum of cleaning products in those supermarket aisles is not only naïve, but also scientifically incorrect.

I became interested in the chemistry of cleaning when I was trying to work out how to unclog my kitchen sink. Prior to that, I'd only ever thought about food in terms of what happens to it in a pot or on my plate, never where it went after I'd thrown it away.

Then again, I realized, it's still food whether it's jamming the pipes of my sink or traveling through my digestive system. I began to think about cleaning in the same way I'd previously approached cooking: as just another branch of household chemistry. It might not be quite as fashionable as *MasterChef*, which was popular at the time (*MasterCleaner* doesn't have the same ring to it), but it's definitely just as interesting from a scientific point of view.

If you look up "blocked sink" or something similar online, whatever search engine you use will return reams of websites all urging you to try out some old-fashioned tactic of yore. A YouTube video on natural remedies racked up more than one million views simply by showing someone unblocking a drain using nothing more than salt, vinegar, and baking soda.

I immediately realized that this didn't make any kind of chemical sense. The person had mixed up half a pound of salt and baking soda, tipped it into the drain until it was full, then poured down boiling water mixed with a liter of vinegar. When vinegar and baking soda combine, carbon dioxide is released, which causes fizzing. This doesn't mean the sink is being unclogged—it's just that the vinegar has reacted with the baking soda, therefore canceling it out.

The person in the video concluded by saying, "I do this weekly," which is when, I confess, I burst out laughing! If he was doing it weekly, then clearly he wasn't unclogging anything. I couldn't stop myself from scrolling through the many comments from irate individuals who'd posted things like, "I took your advice, and now my pipes are so completely jammed, not even a drop of water can get past. What an idiot—I should never have listened to you! I've had to call a plumber!" followed by long lists of insults.

I know we shouldn't laugh at people's misfortune, but I couldn't help it. I'm fed up with seeing people suggest the ridiculous combination of vinegar and baking soda (in this case, with table salt thrown in for good measure) in places where it won't help at all. It was this pet peeve that got me thinking, "Should I—could I—write an explanation of the science of cleaning?" Well, I did, and here it is.

1

Clean and Dirty

CLEAN

Clean, wash, deterge, disinfect, sterilize, sanitize: We might think we're already familiar with most of these words. However, despite their similar subject matter, they don't all mean the same thing, and misunderstanding them can lead to mix-ups and misuse of cleaning products. For example, baking soda neither disinfects nor cleans. Bleach, on the other hand, can be used to disinfect but not clean. Soap cleans and sanitizes but can't disinfect or sterilize. Before we delve into the science of cleaning, we need to agree on the precise meaning of the words we're going to use—such as the distinction between "sanitize" and "disinfect," which were often used interchangeably during the early years of the COVID-19 pandemic. There may be no universally agreed-upon definitions, but it does make sense to clarify how they'll be used in this book.

Clean

To clean means to remove dirt or unwanted matter, which is often visible but can also be invisible. A variety of mechanical means, detergents, chemical solvents, or other methods can be used to dislodge or disperse dirt. Often, a combination of methods is adopted. (We also have to be clear about what we mean by "dirt," but we'll deal with that later.)

In everyday life, "cleaning" is a generic term that encompasses a variety of activities, from sucking up dust off the floor to removing tarnish from silver jewelry to washing or scraping food off plates and pots. I put the word "cleaning" in the title of this book for precisely this purpose: to show how vast a subject it is.

Wash

This term refers to cleaning a solid surface with a liquid, usually water (although dry cleaners use water-free liquid solvents, and we at home might use organic solvents—anything from acetone to ethanol). Sweeping the floor is cleaning, whereas mopping the floor with soap and water is washing. Even using water alone with no detergent is still washing.

The cleaning power of water is often underestimated. Water and a cloth are generally more than enough to tackle a layer of dust: The dust will stick to the cloth, removing it from the tiles or table, although a detergent could also be used for a better all-around clean.

Deterge

You've likely encountered this uncommon term only as part of the more everyday word "detergent." What's a detergent? The name might make you think of laundry detergent, but detergents can come in solid or liquid forms and are used for many purposes. According to the International Union of Pure and Applied Chemistry, a detergent is "a surfactant (or a mixture containing one or more surfactants) having cleaning properties in dilute solutions."[1] Later on, we'll attempt to clarify what exactly a surfactant is. For the time being, my main purpose is to draw your attention to the fact that a variety of substances commonly recommended for household cleaning are not actually detergents.

As we just saw, "detergent" is often used to describe a formulation that contains a combination of surfactants and other ingredients. A formulation is a specific mixture of ingredients in a product, each one performing a specific function—whether it's giving the product a fragrance, keeping it stable over time, preventing the proliferation of microbes, maintaining its pH, or something else. Formulators, who are responsible for creating this mixture, play an important role in both cosmetics and cleaning industries.

At this point, establishing what it means "to deterge" is easy: to clean using a detergent. Wiping the table with water is not deterging—it's merely washing. This doesn't mean you

can't complete some cleaning tasks with water, but to deterge (that is, to clean thoroughly), you have to get a detergent involved. That tomato sauce stain on your shirt won't go away with water alone; it'll need a detergent to shift it. In everyday lingo, we often mean "clean with a detergent" when we say "wash." As long as we all know that, we can avoid any misunderstandings.

Focusing purely on the visible outcome of cleaning runs the risk of oversimplifying. Using a detergent often does much more than simply removing the visible dirt (like a stain on your shirt or leftover pasta sauce on a plate), and it may even get rid of invisible bacteria that could be harmful to our health. The effectiveness of cleaning with a detergent depends on a number of factors: mechanical action (for example, scrubbing), chemical action (the detergent itself), and the temperature and duration of the intervention. In order to disinfect and sterilize, cleaning with a detergent must take place first because dirt is rich in microorganisms that will actively multiply and inhibit the action of disinfectants.

---

1 Alan D. McNaught and Andrew Wilkinson, eds. *Compendium of Chemical Terminology*, 2nd ed. (Oxford: Blackwell Scientific Publications, 1997).

Improvements in hygiene over the past few centuries, primarily thanks to wider availability of detergents (especially soap), have undoubtedly contributed to increased life expectancy in the developed world. To maintain sufficient standards of personal hygiene—to clean the clothes we wear and most of the surfaces in our homes—washing and cleaning with detergent more than suffices.

Deep clean

The concept of deep cleaning (referred to as "hygienic cleaning" or "hygienizing" in some countries) has crept into the cleaning industry to the extent that in a number of European countries, there are even specific words (and a category of cleaning products!) for it. Its actual meaning, though, can be confusing. The closest definition might be to clean deeply in order to "make hygienic," in the sense that removing harmful substances potentially present in the home environment restores it to a safe level of hygiene. This does not go as far as disinfecting, however, which takes the removal of harmful substances to the next level. So, strictly speaking, a "deep or hygienic cleaner" just cleans with a strong detergent that removes some, but not all, bacteria and other dirt.

Disinfect

Disinfecting is the almost total elimination of harmful microorganisms, including bacteria and viruses that can cause infection and disease. With some important exceptions (including if you have a weakened immune system) there is no need to disinfect your clothes, since the bacteria normally present in our homes and on our clothes and skin don't have to be killed. Our skin is home to millions of microorganisms (commonly known as the microbiota), which perform vital functions just like the friendly bacteria in our gut that we couldn't survive without. Constant disinfection of everything around us is unnecessary—in most cases, simply washing and removing dirt will do. There are times, though, when disinfecting is required.

I'll say more about that later, but one thing I'd like to get straight here is that disinfecting requires the use of specific biocide or medical/hospital-grade disinfectant products. These products' abilities to remove microorganisms (whether viruses, bacteria, or fungi) must be clearly stated on the label and backed up with scientific evidence to comply with government regulations. (So, no, baking soda is not a disinfectant, either.)

Sanitize

"Sanitize" is related to "disinfect," but not directly equivalent. Products that are marketed as sanitizers can remove bacteria from surfaces, but not viruses or other microorganisms.

DIRTY

The stuff around us (like clothes, table linens, and bedding) and the various surfaces in our home (such as floors, dishes, and other objects) are the landing places for a variety of unwanted impurities, otherwise known as dirt. Ideally, we want to remove that dirt while leaving the objects themselves intact.

The first way dirt can be classified is whether it's soluble in water. Grains of sugar and salt, for example, are water-soluble, meaning they can be easily removed with (preferably warm) water. Insoluble material is generally oily or consists of matter like clay or soot.

A second way to categorize dirt is by its physical state: either liquid (a drop of oil that ends up on the tablecloth rather than on the ribs and side salad you just made) or solid (the thick grease left on the napkins from said ribs, or coal from the barbecue that got onto your pants while you were grilling). Some dirt falls somewhere between liquid and solid. In fact, many of the things we use are half-solid, half-liquid—like jam or mayonnaise.

Generally speaking, insoluble solids are the most difficult to remove. In some cases, though, solid dirt (remember the fat from the barbecued ribs?) will liquefy when heated and become easier to clean. That's why you should always think about the temperature of the water you use when trying to clean something.

Most of the time, dirt has a color, and the stains we typically talk about the most are the highly visible ones from wine, coffee, ink, or sauce. These have vibrant colors and are therefore easier to spot no matter how small they are. They don't usually wash away with ordinary detergents. A stain may have been made by something that was water-soluble (a marker pen, for example), but the color molecules have a nasty habit of remaining stubbornly attached to the fabric fibers and withstanding any kind of cleaning with a detergent. Only attacking the stain with chemicals will make a difference (which we'll come to later).

If you think about the many types of dirt and the different properties of each, it's no surprise that cleaning is such a hugely complicated business that needs to be addressed on a case-by-case basis. When we talk about oily

DIRT CLASSIFICATIONS

LIQUID	HALF-SOLID, HALF-LIQUID	SOLID
Oil	Jam	Thick grease
	Mayonnaise	Coal

substances, for example, we could be talking about anything that is not soluble in water and is greasy to touch. Edible fats (oils that are liquid at room temperature are still considered fats) clearly fall into this category, along with fatty acids (some of which have a very unpleasant smell), plus a variety of alcohols, hydrocarbons, waxes, and much more.

A common form of dirt that builds up on our clothes is the mostly fatty material shed from our skin. Sweat also secretes both soluble substances (like sodium chloride and urea) and insoluble ones (which can include fats, proteins, and pigments). The fatty parts are mostly sebum secreted by our sebaceous glands as well as a small percentage of fatty material from dead skin cells. Sebum is solid at 50°F (10°C) but gradually softens the closer it gets to our body temperature of 98.6°F (37°C), by which time it is fully liquid. Knowing this, if you ever want to wash something with sebum on it, make sure you do it at 100°F (38°C) or higher.

Sebum is made of a complex mix of triglycerides, a bit like edible fats, but it also contains mono- and diglycerides, esters, squalene, and other substances. This complex composition makes it difficult to get rid of compared to a splash of oil on your shirt at lunch. To make things even more complicated, once sebum gets inside a fabric's fibers, it reacts chemically. The free fatty acids and squalene combine to form a polymer, which is the yellow substance that stains your clothes and bedding. Sebum also contains suspended solids that come from dead cells of dry skin.

Non-oily solid dirt is equally difficult to budge, especially if the molecules have a chemical structure that enables them to bind with the molecules of the textile fibers in a more or less stable way. This might be a result of polarity, shape, surface properties, or a multitude of other parameters. Common solids present in dirt include lime residues, the carbon products of combustion (have you ever looked at the underside of a pizza cooked in a wood-fired oven?), dust, sand, cellulose, and much, much more.

This rapid but absolutely not exhaustive roundup of types of dirt has hopefully shed some light on why supermarkets sell so many different types of cleaning products. Different kinds of dirt, deposited on a variety of different surfaces, need equally diverse methods and products to remove them. Floors are different from shirts, just as pasta sauce has little in common with tarnish on jewelry. But before you can pick the right product and method, you need to understand the science underpinning them.

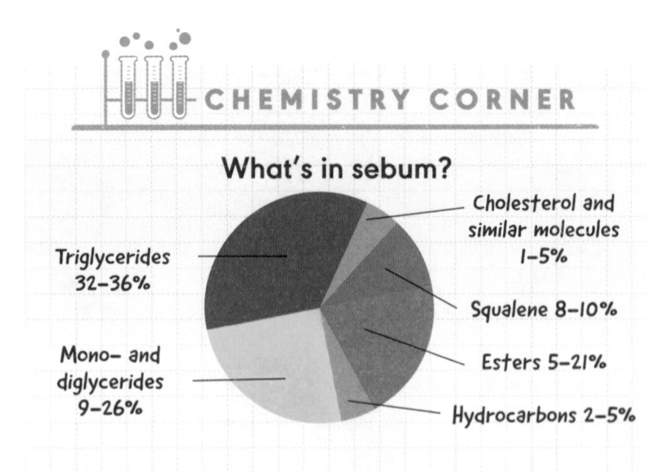

HAZARDS AND RISKS

Before we draw this chapter to a close, there's one more thing I'd like to address: the difference between hazards and risks. The two concepts are not the same, and mixing them up can sometimes cause us to underestimate an actual risk and expose ourselves to it unnecessarily or worry too much about one that is practically nonexistent.

If you read the labels of many common household cleaning products, you'll find an array of caution statements and notices about hazards to humans, animals, and the environment. But how worried should we actually be about our safety around these products? One of the premises of safety legislation (which also applies to the food we eat) is that if a product has been authorized for sale, then it should, in theory, be safe to use. How does that work? This is where a clearer understanding of risks and hazards comes in.

A hazard is an object, substance, or situation with the potential to cause harm to a person or the environment. This potential has no connection to external factors. Sharks, for example, are intrinsically hazardous animals, and so are lightning, driving a car, and detergents. For each of these objects or situations, you can easily imagine the damage they could do: being attacked by a shark, struck by lightning, crashing the car, or ingesting a detergent.

How worried should we be about sharks, lighting, cars, and detergents? Well, that depends on the risk we expose ourselves to. If we're out swimming and spot a shark fin nearby, then we'd be right to be very concerned. But if we spot the shark fin from up on a cliff as we gaze out at the ocean, then we can pretty much relax. A shark is a hazard. Swimming beside one is a risk. Standing on the cliff is safe. Driving on the wrong side of the road is a much bigger risk than driving on the right side. When it comes to lightning, it's much riskier to stand under a tree during a storm than to stay inside. As another example, an egg is hazardous because it could contain salmonella. Eating a raw egg is riskier than cooking the egg before eating it.

To put it another way, risk is the likelihood that an object, substance, or situation will cause harm. It's a probability, a number, and it depends on how much we're exposed to the risky thing. Unfortunately, we humans are not very good at understanding probabilities. Often, we'll become alarmed over a hazard and disregard the actual risk. Risks need to be objectively assessed to be effectively reduced. This is exactly what lawmaking bodies do: Assess risks, then suggest steps people can take to make sure the products they buy—whether those are medicine, food, cosmetics, or detergents—are safe to use.

Aspirin, descalers, and bleach are all hazardous. For this reason, the relative hazard is clearly indicated on the information leaflet included with medications and marked on the bottles, boxes, and containers of cleaning products. That said, if we then use them correctly, the risk of anything untoward happening is minimal. It will never be zero, unfortunately. Something could always go wrong where humans are involved: A product might spurt in someone's eye, or the wrong dosage of medicine could be taken.

As new studies come along and new information is learned, the risks associated with such hazards are reassessed regularly. This means a previously approved chemical might be withdrawn from the market and replaced with one that is less dangerous or brings much fewer risks.

2

Acids and Bases

I was in third grade when my parents gave me a young scientist's kit. That kit was the spark that ignited a lifelong passion for all things chemical and shaped my choices, first at school and then professionally. It was a real box of delights, complete with test tubes, glass flasks, and an alcohol burner as well as an array of reagents in the form of multicolored salts: copper sulfate, potassium permanganate, sodium dichromate, nickel chloride, and iron sulfate.

Many of these are no longer included in modern-day kits, and glassware (never mind an alcohol burner) is now just a figment of the imagination. Today, it is generally accepted that anything made for kids has to be one hundred percent benign, meaning modern "young scientists' kits" no longer contain anything of any interest. Most of the wildly exciting experiments that fueled my passion for chemistry would be impossible these days. If you can't heat anything up, then what you can do is kind of limited—all you have are a bunch of plastic containers, some salt, some baking soda, and a few minor reagents, all of which are harmless and therefore completely useless. Nowadays, I'd honestly be surprised if a young person fell in love with science after playing with one of these kits.

There may not have been quite the same focus on safety back then, but the old junior science kits still lacked some of the key reagents needed to do proper chemistry. Mine had no basic substances and no strong acids, both of which are admittedly fairly hazardous. But grown-up chemists must at some point learn to handle them safely in the same way a chef has to work with boiling water and learn to deep-fry. I managed to source some sodium hydroxide (more commonly known as lye) from a local corner store, but acids were more difficult to come by.

One of my experiments required dilute sulfuric acid, which I knew was found in car batteries. My dad kept an old battery in the garage, so I brought it inside one day, thinking I could get the acid out of it. Back then, it was much easier to get a car battery open. But as I poured out the acid, a little bit splashed onto my mom's best rug. (Yes, I was enough of an idiot to be doing this in the living room.) At that age, I had no idea how corrosive sulfuric acid was, but I soon learned: The very next day, Mom found a hole the size of a tennis ball in the rug! The acid had literally dissolved the fabric, disintegrating it and leaving a charred edge as if it had been burned. From that day on, all my experiments were confined to my desk in the attic, and I was very careful when I explored the properties of acids and bases.

Both acids and bases (the name we chemists give to alkaline substances) can burn through fabric like my mom's rug, but in weaker concentrations, they can also be highly effective at cleaning it. When it comes to cleaning and chemicals, acidity and alkalinity are probably the most important concepts of all in both theory and practice—namely, in knowing how to pick the right product for the job at hand.

DID YOU KNOW?

The opposite of an acid is a base. Substances (or solutions) are therefore described as either acidic or basic. For example, acetic and hydrochloric acid are acidic, while sodium hydroxide and baking soda are basic.

THE pH SCALE

Most people know that pure water has a pH of 7 and that this is considered neutral. Dissolving other substances in water can change a neutral pH and make the resulting solution acidic (with a pH lower than 7) or basic (with a pH higher than 7). Some acids, like hydrochloric or sulfuric, are classified as "strong" because high concentrations will decrease the pH of a solution to very low levels, sometimes even to 0. Others are classified as "weak" because, regardless of their concentration, they don't lower the pH by a great amount. A typical solution of a weak acid will have a pH between 2 and 5. At the other end of the scale are bases. Like acids, "strong" bases such as lye can raise the pH to higher than 13, whereas "weak" bases like baking soda will raise it by only a few degrees, typically settling at around 8 or 9. Needless to say, these are only general definitions—pH fluctuates constantly between 0 and 14.[1]

For our purposes, we don't need to go into too much detail about how pH is determined mathematically or measured experimentally. All we need to know is that it is connected to the number of H^+ (hydrogen) ions in a solution: The lower the pH, the more hydrogen ions it contains.[2]

Strictly speaking, pH always refers to a solution of a substance in water. This means that when we talk about the pH of baking soda or citric acid, for example, both of which are solids, we are actually referring to a water-based solution of these substances at a particular concentration—say, 10 grams dissolved in one liter of water. The pH of the

pH MEASURING DEVICES

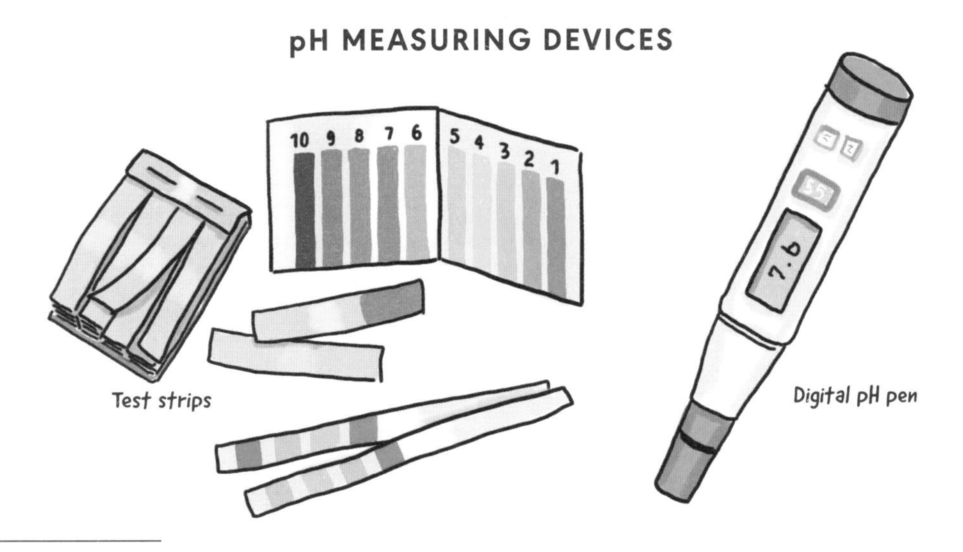

1 pH values below 0 or higher than 14 are also possible, but not in any situation we might encounter in the home or with domestic cleaning products.

2 In its simplest definition, pH is equal to the negative logarithmic activity of hydrogen ions in solution.

resulting solution depends on both the substance itself and the amount of it dissolved.[3] If a very small quantity of an acid or base is dissolved in water, the pH will change only by a very small amount. If larger quantities of a substance are used, the pH will be much closer to 0 (if it is acidic) or 14 (if it is basic). That said, in most cases, there is a pH value beyond which a solution will not go. Baking soda, for example, is a basic substance, but when it is dissolved in water, the pH will not move beyond 9, even in the largest possible concentration.[4]

You can easily measure any liquid's pH using a test strip soaked with a universal indicator, which is a chemical compound that changes color to indicate (hence its name) the approximate pH of the solution. Alternatively, you could also try a liquid kit used to measure the pH of aquarium water. If it's a precise measurement you're after, maybe for making homemade soap, then you'll definitely need a digital pH pen, which is easily purchased online.

7 is the exact pH of distilled, impurity-free water. Domestic tap water rarely has a pH of 7—it fluctuates between 6.5 and 8.5 depending on the dissolved minerals and carbon dioxide. Water that's treated before it enters our homes tends to be given a pH of 8 to prevent any potential corrosion of pipes. (By the way, did you know that carbonated water has a pH between 5 and 6 due to the dissolved carbon dioxide?)

THE pH SCALE RAINBOW

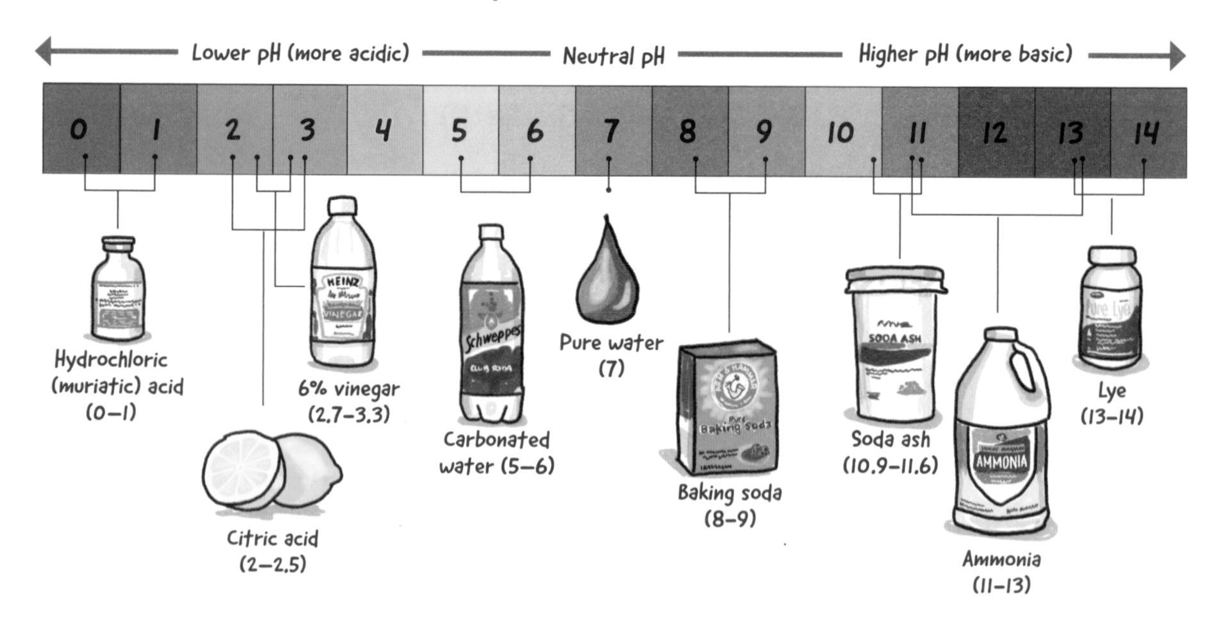

3 Other factors like temperature can also affect the pH of a solution, but these can be overlooked for our purposes.

4 The maximum amount of baking soda that will dissolve in one liter of pure water at 68°F (20°C) is 100 grams.

ACIDS

The most acidic substance that's still occasionally used for specific types of cleaning at home is hydrochloric acid, sometimes called muriatic acid. Concentrated solutions of this acid have a pH of 0 to 1 and are highly corrosive. Less common but just as corrosive is sulfuric acid. Depending on the strength used, these acids can be extremely effective at removing residues, especially those with the opposite basic nature. The most obvious examples are limescale, also known as calcium carbonate, and rust, which is a mixture of iron oxide, carbonate, and hydroxide.

Strong acids should be handled with extreme care. Most importantly, make sure you wear gloves and safety glasses at all times when using them because the tiniest splash on your skin or in your eye can cause enormous damage. They're also too corrosive for most of the things we want to clean. Remember my mom's rug? Hydrochloric acid removes limescale from faucets, but it's so strong that it can also damage the metal, which is why much weaker acids are generally preferred. However, it's still used in a number of products designed to descale the crustiest of toilets, since porcelain is much more resistant to acids. Also, sulfuric acid appears in a variety of drain cleaners.

Moving along the pH scale, we now come to weaker acids. Their low pH makes them more useful around the house and a key ingredient in most consumer cleaning products. Citric acid is a prime example. A naturally occurring acid found in citrus fruits, it is also the preferred organic acid used in commercial cleaning supplies. A citric acid solution has a pH of 2 to 2.5, similar to natural lime or lemon juice, and is a popular ingredient in descalers (and a variety of other products, as we will see).

Further along the scale, a little less acidic with a slightly higher pH, is ordinary vinegar, which gets its acidity primarily from acetic acid. An ordinary 6 percent–strength vinegar measures around 2.7 to 3.3 on the pH scale. Vinegar is also used in domestic cleaning—although not always properly, given that it can't actually do some of the great things ascribed to it. It won't stop limescale from building up in your washing machine, and it is neither a degreaser nor a disinfectant. Unlike citric acid, it is not widely used in

EFFECTIVE ON
Limescale deposits
Urine stains
Dried soap and associated calcium salts
Concrete
Thin layers of corrosion on metals (like rust and verdigris)

DAMAGE
Limestones (like marble, dolomite, travertine, and alabaster)
PVC
Metals
Fabrics (especially cotton)
Coral, pearls, and shells

commercial cleaning products for several reasons: On top of its acrid smell, vinegar is highly corrosive to metal (copper, bronze, and brass in particular), damages rubber seals in household appliances, and strips away the coating on tiles and grout (not to mention the joints between them).

Other organic compounds used in cleaning products include lactic acid, malic acid, and formic acid. Phosphoric acid (present in a variety of fizzy drinks like Coca-Cola) is often used to descale faucets and sinks because, unlike hydrochloric acid, it doesn't damage chrome plating.

Moving even further along the pH scale, we come to most food and drinks, the majority of which are almost never used as cleaners—primarily because they are generally the cause of the very stains we find ourselves cleaning or the incrustations that require something stronger to dislodge them.

DID YOU KNOW?

Our stomach contains hydrochloric acid at a pH between 1 and 2 to start digesting food.

BASES

Switching to bases used for cleaning, the first one we encounter is baking soda, a favorite in most homes.[5] However, just like vinegar, it is often used in the wrong way. It is much less popular among store-bought cleaning products, as people typically purchase more alkaline alternatives. A baking soda solution has a pH that's between 8 and 9 and is unlikely to get much higher, as it's poorly soluble in water. In comparison, the pH of common soap is between 9 and 11, and it's a relatively powerful cleansing agent. That said, it can also be an irritant when the highly basic pH of some soaps meets the acidic pH of the outer layer and several of the inner protective layers of our skin. This explains, in part, why soaps have been almost entirely replaced by synthetic detergents with much lower pH values. For the very same reason, most shampoos have an acidic pH, which is much better for your hair than soap.

Generally speaking, a basic pH helps remove surface dirt, especially from cotton fabrics. This is because both the cotton fibers and most types of dirt are negatively charged (that is, these materials contain more negatively charged electrons than positively charged protons). Substances with the same charge repel

DAMAGE
Painted surfaces
Linoleum
Aluminum
Wool and silk

5 Also known as bicarbonate of soda or sodium hydrogen carbonate, it has the chemical formula $NaHCO_3$.

each other (while substances with different charges stick together more easily), and a basic pH increases the repulsion between the two, helping the action of the detergent. Additionally, at a basic pH, proteins and starches bond with water much faster and fats are partially converted to soap. But we will get into that shortly.

At higher pH levels, we have a range of chemical substances used to clean the home and even more employed in mass-produced cleaning products, most of which are basic. The oldest is undoubtedly sodium carbonate, more commonly known as soda ash.[6] Before the invention of soap, soda ash was used for washing, a practice that did not fade away entirely even after the evolution of soap and can still be found today. A soda ash solution cleans most effectively at a pH of up to 11.5.

6 The chemical formula for sodium carbonate is Na_2CO_3.

While everyone knows that acids can be corrosive and dangerous, there is much less awareness about what high-pH basic substances can do. They are equally corrosive and able to remove a variety of types of dirt just as effectively. To safeguard ourselves and the materials we want to clean, we should be just as wary of bases as we are of acids. Bleach is a basic substance with a pH somewhere between 11 and 13 that is so corrosive in high concentrations, it damages faucets and discolors chrome. This is not even due to its pH, which is not excessively high, but to the presence of the powerful oxidizer sodium hypochlorite.

Another basic substance, commonly used in cleaning products primarily for its solvent properties, is ammonia. Strictly speaking, ammonia is a gas that is dissolved in water in varying concentrations and sold as a liquid. Its pH ranges from 10.5 to 12.

The most basic substance regularly used for cleaning is sodium hydroxide, popularly known as lye. It was once used to clean heavy fabrics like canvases and sheets. Pure lye is sold in flakes stored in airtight plastic containers. A highly corrosive and hazardous substance, it should be handled with the utmost care. You can find it in a wide range of products, either as a source of alkalinity or, in higher concentrations with a pH over 13, as a way to unblock drains.

DID YOU KNOW?

Almost all food has an acidic or near-neutral pH. Egg whites are the only food with a basic pH higher than 9.

WHAT'S THAT SUBSTANCE?

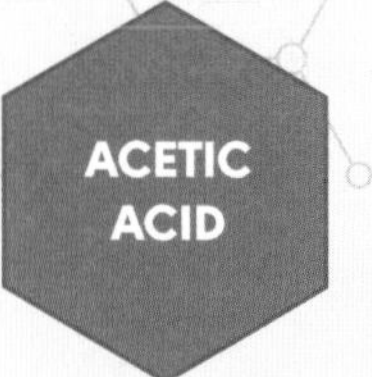

Pure acetic acid is a clear, colorless liquid at room temperature. It has a characteristic pungent odor and is also referred to as "glacial acetic acid" because of the way it solidifies into ice-like crystals at 61.9°F (16.6°C). Aside from the obvious vinegars and pickled vegetables we're used to seeing it in, solutions of acetic acid occur in many other foods. They are also used as food additives (called E260 on food labels) that regulate acidity or act as a preservative. Some medications and cosmetics contain acetic acid, as well as pesticides and cleaning products—although always mixed with other ingredients. It is added to pet shampoos and used as a solvent in silicon sealants, and it has applications as an agricultural herbicide that controls weeds.

Acetic acid is widely used as a raw material in the industrial production of many substances, from synthetic resins like polyvinyl acetate (PVA) to semi-synthetic cellulose acetate. It's an excellent solvent, dissolving water-soluble substances like salt and sugar as well as other organic materials that are nonpolar—that is, made of molecules whose electrical charges are distributed evenly throughout, allowing them to mix easily with other nonpolar substances to form solutions.

Acetic acid solutions used in the food industry are produced by bacteria, while most of the acetic acid required for industrial applications is made chemically usually using methyl alcohol and carbon monoxide.

It was probably discovered that vinegar could be obtained from bacteria in the genus *Acetobacter* around the same time it was learned that alcohol could be produced by fermentation. Indeed, alcohol exposed to oxygen is what enables bacteria to excrete acetic acid. The first reliable accounts of acetic acid production date to the Babylonians five thousand years ago. Not surprisingly, they were the first to develop the fermented alcoholic drink we know as beer—which was also where they got their vinegar from when it didn't come from figs and dates. In contrast, the ancient Greeks and Romans obtained their alcohol primarily from wine, while other cultures used a variety of raw materials—for example, malt in England and sake in Japan.

Vinegar is produced by bacteria, usually by the commonly occurring *Acetobacter* or *Clostridium acetobutylicum*. The source makes no difference as long as it contains ethanol from a previous yeast-induced fermentation. The end result is always acetic acid, along with the malodorous remains of whatever set off the reaction in the first place.

Although vinegar is predominantly associated with wine, it can also be made from apple cider; fermented grains like rice, corn, and barley; fruits like dates and figs; and even the sap of certain plants, like coconut palm or sugar cane. The healthy dose of fermentable sugars in most plants means they can generate an equally large variety of vinegars that contain not only acetic acid, but also the aroma of the original fruit. Raspberries, persimmons, strawberries, and an array of other fruits are turned into aromatic vinegars around the world and sold as artisanal condiments. Gastronomically speaking, they may be imbued with a tempting selection of different flavors and fragrances, but the acetic acid doesn't work any differently for cleaning.

⚠ WARNING!

The chemical substances I've been describing are all highly reactive: They're programmed to bond with other substances to form new compounds. Specifically, acids react easily with basic substances and vice versa. So, if it's limescale (a base) you want to remove, then it's acid you need, whereas the alkalinity of baking soda is ideal for cleaning the acidic residue of tomatoes from the back of the fridge.

The one thing you should never do, though, is mix acids and bases in a cleaning solution. It's not like blending oil and vinegar into a tasty vinaigrette—it starts a full-blown chemical reaction. Generally speaking, chemical substances and cleaning products should never be mixed, randomly or otherwise. Why? First and foremost, because each loses its original chemical properties and becomes something else. Unfortunately, when you dip into forums and Facebook groups about cleaning, it isn't long before you come across someone advising you to do just this and haphazardly mix products together in the misguided belief that you can make something that combines the qualities of each.

That's not how chemistry works, I'm afraid. If you put two cleaning substances together and they react, the end result will be, at best, something that no longer has the properties of the original substances and is nothing short of useless. At worst, the product of the reaction could be extremely dangerous. Every year, hundreds of people end up in the ER with poisoning and lung scarring because they poured a mixture of hydrochloric acid and bleach down the toilet with the mistaken idea that it would double the cleaning power. Tragically, such accidents can sometimes have fatal consequences.

I'll say it again and again: Please do not mix chemicals unless the label specifically instructs you to do so. With each extra warning, I hope there'll be one fewer person heading to the ER.

FIGHTING DIRT WITH ACIDS AND BASES

From the few examples we've encountered so far, it should be clear that no one product can clean everything—and that before you choose a product, you need to think about the type of dirt you want to remove. A product's pH tells you a lot about what it can and can't clean, but that's by no means the only criterion we need to consider. Alas, regardless of whether you choose an acid or base, too many common types of dirt will refuse to budge. That's why so much time and money have been spent over the centuries trying to come up with (or "synthesize," as we chemists would say) better, faster, more effective cleaners, starting with the invention of soap.

Let's go through the dirt-fighting process in stages. Firstly, water alone is sometimes enough. If there's sugar or salt on your tablecloth, once you've scraped off the bulk of it and shaken the tablecloth, washing it in water is enough to dissolve any grains still trapped in the fabric's fibers (given that we're dealing with soluble substances). The same goes for all colorless soluble substances. (Not coffee, obviously, which is soluble but stains due to its dark color.)

The most common type of dirt on the body and in the home is grease. Since grease is a buildup of oily matter, acids struggle to deal with it, as they are polar molecules, which can't dissolve nonpolar substances like oils. If anyone has ever told you to use vinegar or lemon juice on greasy stains, they're wrong! If you don't believe me, try pouring some oil onto a napkin and tell me what happens when you squirt it with vinegar or lemon juice instead of a detergent (which is a base). Unlike weak acids, bases with a high enough pH are very good at removing oily stains—in some cases even turning them into soap to multiply the cleaning power. This explains why most cleaning products are basic, as I mentioned earlier.

Sugar cleans up fine with a splash of water, but the same can't be said for many other substances in its carbohydrate group. For example, starches—found in pasta, bread, and rice—are carbohydrates but not sugars. More specifically, these foods contain complex carbohydrates, which are insoluble in water. We couldn't boil spaghetti for dinner if they were soluble! Many vegetables, like potatoes, grains, or pulses, are also starchy and can be among the most difficult substances to remove. They won't budge with either acids or bases,[7] while detergents can only facilitate the mechanical action of dishwasher spray arms or human hands struggling to physically remove starches from the materials they've penetrated.

After grease and carbohydrates, we'll end this section with another group of substances often found in dirt: proteins. Strong-arm

7 With the exception of strong acids like sulfuric acid, which should definitely not be used to wash the dishes.

tactics deploying both acids and bases often succeed in removing proteins because they succumb to strong acids and strong bases. This is why products containing lye successfully dissolve food proteins (or any type of hair) clogging drains. Sulfuric acid also works, but only as a last resort because it doesn't stop at dissolving the hair: It also breaks down grease, protein, carbohydrates, and a whole list of other things. (If safer methods have failed, my advice is to please, call a plumber instead of resorting to sulfuric acid.)

One last thing before we move on: Dirt changes with the temperature. In most cases, moderate heat can make cleaning easier, as things like butter or other saturated fats that are solid at room temperature melt when heated and wash away much faster with a detergent. That said, if the temperature rises too much (maybe to that of an oven), some fats polymerize and become almost impossible to remove. You may have noticed this when trying to clean the caked-on brown marks inside the oven's glass door and in the oven itself.

Sugars dissolve easily in water, but when heated, they caramelize and become much trickier to clean. Proteins also denature and coagulate when heated, making them equally hard to remove. For this reason, blood should be cleaned with cold (not hot) water to stop the protein from coagulating and getting "cooked" into the fabric.

Acids, bases, and soap can't clean everything. So, the next chapter will look at how another group of chemicals—enzymes—have come to the rescue of detergents over the past few decades, helping them tackle greasy stains or say goodbye to unwanted protein and starch.

REMOVING DIFFERENT TYPES OF DIRT

COMPONENT	SOLUBILITY	REMOVAL	EFFECT OF HEAT
Sugars	Soluble in water	Very easy	Caramelize, making them more difficult to clean
Starches	Insoluble in water; not very soluble in acids and bases	Difficult	No significant change
Protein	Insoluble in water; a little soluble in acids, soluble in strong bases	Very difficult	Denature and coagulate, making them more difficult to clean
Fat	Insoluble in water and acids; soluble in strong bases	Difficult	Melt in moderate heat to become easier to clean, polymerize at high temperatures and become much more difficult to clean.
Inorganic salts and oxides	In water, varies from highly soluble (table salt or sodium chloride) to more or less insoluble (rust or iron oxide); usually soluble in acids	Varying difficulty	No significant change

Baking soda and vinegar: a mixture that makes no sense

Every time someone suggests mixing vinegar and baking soda to remove a stain or unblock a drain, a chemist somewhere combusts. I couldn't say precisely when this particular craze began, but it's definitely popular. Pick any online cleaning forum or Facebook group, then look for a video on how to unblock the sink, clean the carpet, or degrease a frying pan—or follow a few influencers or pick through the handy hints section of a modern home magazine—and it won't be long before you begin to hear the inexorable chant in your head: "Vinegar and baking soda, vinegar and baking soda, mix them quickly and watch the magic."

It's a pity that not only does mixing them not work, it can even be counterproductive. I know many of you are thinking, "But everyone is saying it!" Well, I'm afraid that everyone is wrong. No matter how many times we repeat something false, it doesn't become true. As I explained above, baking soda is basic while vinegar is acidic. When mixed, they react instantly to produce water, carbon dioxide, and sodium acetate, a mildly basic substance with absolutely no cleaning properties. Likewise, if you mix—in the correct quantities—two highly corrosive substances like lye and hydrochloric acid, the result is an innocuous water and sodium chloride mixture: salt water, in other words. This happens because in chemical reactions, the properties of the original substances disappear as the substances themselves no longer exist, having chemically transformed. Therefore, it makes absolutely no sense to mix substances that react with each other.[8]

I realize, however, that it might be too flippant to dismiss vinegar and baking soda in this way. As a chemist, I've often wondered how such an inaccurate piece of advice could have become so popular. It wouldn't be the first time a wholly incorrect or ineffective home or traditional remedy has propagated at such speed and with such fervor. Kitchen and cooking tips are an excellent case in point, and some of the popular claims about how to clean or keep our kitchens hygienic are so wildly inaccurate that they shouldn't be given the time of day. Why would throwing coffee grounds down the sink help unblock it?

Anyway, the habit of mixing vinegar and baking soda is so common, and its promoters are so convinced of its effectiveness, that I decided to take a closer look. I spent a long time thinking about it until I eventually identified three reasons this urban legend has taken such a strong hold.

The first is psychological: When vinegar (or lemon juice) touches baking soda, it

8 Except in a small number of extremely rare cases where it's explicitly done to trigger specific chemical reactions.

WHAT'S THAT SUBSTANCE?

SPIRIT VINEGAR

If instead of fermenting wine or another alcoholic beverage, you start the acetic fermentation process with ethyl alcohol (also called ethanol) and dilute it with water in the correct proportion, you get spirit (or alcohol) vinegar, a solution of acetic acid in water. This perfectly clear, colorless liquid is sold in supermarkets alongside the rest of the vinegar, generally with a lower price tag. The acetic acid in spirit vinegar is exactly the same as that in other vinegars. So, while it may lack the aromas and flavors acquired from grapes, apples, or other sources, making it less attractive for culinary purposes, it is ideal for cleaning. Why? The water and acetic acid are completely volatile (meaning they evaporate easily), so they don't leave a stain.

instantly fizzes, producing a very impressive cloud of foam and bubbles. It may look like something special is happening, but it's still just carbon dioxide with no detergent properties. If you put it in the kitchen sink, the bubbles may drag up some of the dirt from the pipes and you may interpret this as a cleaning miracle. I'm sorry to be the bearer of bad news: It isn't.

The second reason is the comforting knowledge that both vinegar and baking soda are edible, so they must be completely harmless. Everywhere we turn, we're bombarded with warnings and ads fueling concerns for our health and the environment, suggesting—incorrectly and often dishonestly—that normal cleaning products, the ones "made of stuff with complicated names we don't understand," are hazardous for our health. Conversely, there's no denying that if you buy a product to unblock the kitchen sink, it will have a warning symbol blazoned across the label. For example, as I said earlier, lye should be used with caution, but it's precisely its corrosiveness that enables it to unclog your sink. So if you pick up a magazine or follow an influencer and both tell you to use a much safer mix of vinegar and baking soda instead of the more dangerous lye, the temptation to believe is difficult to resist, especially if you've forgotten the chemistry you learned in school.

I've no doubt that many of you reading this will be ready to swear that your mixture "worked"—that the last time you used it, it really did clean the thing you set out to clean. I believe you, I do, but let me tell you why.

This brings me to my third reason for this phenomenon, which is strictly chemical. I said earlier that vinegar and baking soda (just like lye and hydrochloric acid) cancel each other out, but only if you use the correct quantities of the two reactants. I'll save you the calculations, but a liter of ordinary 6-percent vinegar[9] requires exactly 84 grams of baking soda to react fully and produce a solution of water, carbon dioxide, and sodium acetate. Or, if you prefer, 100 ml of vinegar reacts fully with 8.4

9 Meaning that 100 ml of liquid contains 6 grams of acetic acid.

WHAT'S THAT SUBSTANCE?

Lye, a basic solution obtained by boiling ash in water, was originally used for cleaning or making soap.[10] Like many other ancient products, its exact chemical composition was not well-defined. Throughout history, lye was a sodium and potassium carbonate solution extracted from ash and boiled down to make a concentrate. However, it could also contain sodium or potassium hydroxide if it was treated with calcium hydroxide (so-called slaked lime) to make it more caustic.

grams of baking soda. After the reaction, both the baking soda and the acetic acid originally present in the vinegar no longer exist. I'm pretty certain none of the concoctions touted as the remedy of all cleaning ills contain exactly these quantities, which is key to understanding why this mixture is believed to work miracles. If you mix less than 8.4 g of baking soda with 100 ml of vinegar, the baking soda completely disappears when the two substances finish reacting, and all that is left is the excess acetic acid that didn't react. Vice versa, if you add more baking soda, the acetic acid disappears and excess baking soda is left. It is the addition of the leftover, unused reactants that makes the mixture look like it is working.

There are generally two types of recipes for mixing vinegar with baking soda: those with an overabundance of vinegar, which create a watery solution, and those with an overabundance of baking soda, where the latter is barely wetted by the vinegar to form a paste. When the leftover reactant is acetic acid, it is still active to a degree against any limescale crusting up faucets, lining pipes, or stopping water from draining properly. This is why the mixture seems to work, even if you're using only what was left after the reaction and not the full 100 ml you started out with. You're wasting vinegar and baking soda to create a liquid that is much less effective.

DID YOU KNOW?

The word "alkali" denotes basic or alkaline substances—those with a pH above 7. It originated from the Arabic "al-qaly," which means "ash," due to the ash used to produce the alkaline substances employed in making soap.

Some people even recommend making the mixture in advance and keeping it in a bottle. I hope you're beginning to see why this makes absolutely no sense: As soon as baking soda and vinegar touch, at least one of them ceases to exist. If any cleaning is being done, it is coming from the leftover vinegar.

In the other recipes, the cleaning power comes from the abrasive properties of the baking soda, which is useful for scrubbing a crusted frying pan or removing buildup from an oven tray. Here, the amount of vinegar recommended leaves a portion of the baking soda unreacted and lightly moistened, so it can then be used to scrape off the dirt. Once again, you'd be better off not wasting vinegar at all and just dampening a small amount of baking soda with water.

10 The word "lye" comes from the Old English "léag," which is also related to the word "lather."

Before soap: washing with ash

The ancient Romans were famous for, among other things, building baths and spas in any territory they conquered. Their ablutions, however, were not done with soap—which was described for the first time by Pliny the Elder (23–79 AD) when he mentioned how the Phoenicians made it from plant ash and goat fat. Greeks and Romans instead washed themselves with water, then with oil and ground lentils. Water is a powerful detergent, but they also used ivory, bone, or metal instruments to scrape any persistent dirt off their skin. Soap, on the other hand, was used for producing a hair-styling pomade.

The ancient Egyptians, however, had already discovered the skin-cleaning benefits of soda ash, as had the Sumerians. Natron, the name they gave to a mixture of sodium carbonate (soda ash) and sodium bicarbonate (baking soda), was extracted from lake water through evaporation, and they used it to clean grease from fabric. As we'll learn, natron was a necessary step that led to the industrial production of soap. But the most popular ingredient used to keep clean in the ancient world was widely—and also freely, given that it was a waste product—available. What was it? Plain old ash from burnt wood or vegetables.

Ash was widely used in ancient Babylonia and across the Arabian Peninsula. Dried and burned samphire, a plant rich in alkaline substances, was the ingredient of choice. Samphire ash is full of sodium carbonate, while wood ash contains both sodium and potassium carbonate.

The plants were burned, the ash was placed in a basket, and boiling water was poured over it. The resulting sodium- or potassium carbonate–rich solution was then used to launder clothes. In royal courts, however, a more complex procedure was followed: The basic solution evaporated and the remaining powder was heated to a high temperature to eliminate water and any remaining organic matter. Then, the powder dissolved when the substance was used.

The use of ash to launder bedclothes and linens (crudely, may I add) continued until relatively recently—around the early 1900s, at least for those unable to afford soap or make it themselves.

In today's world, we tend to romanticize the good old days (which, when you take off the rose-colored glasses, were not as great as they seem), so you won't be surprised when I tell you I've already encountered several green influencers celebrating the use of ash to do laundry, claiming that "ash is natural and chemical cleaners are not!" Don't get me wrong—as a chemist, there's nothing I like more than messing around with chemicals at home. My kitchen and laundry room are often the lab settings for some of my weirdest, most wonderful experiments, like trying to extract basic substances from ash to make cleaning products. It's fun to try once, but I would never suggest it as a serious alternative to a well-formulated detergent.

DID YOU KNOW?

In the Indian subcontinent, Central Africa, the Arabian Peninsula, and other places around the world, ash was once mixed with earth or clay to wash people's bodies and clothes.

WHAT'S THAT SUBSTANCE?

Sodium carbonate (also known as soda ash, washing soda, and soda crystals) is a white water-soluble compound with a variety of uses, from the production of soap and other cleaning products to the manufacturing of food, textiles, and glass. Historically, its impure form was extracted from the ashes of plants grown in sodium-rich soils or from the mineral deposits formed when seasonal lakes evaporate. It was used by the ancient Egyptians for the preparation of mummies because the alkalinity of the soda inhibited the growth of bacteria that would have decomposed the mummies' bodies. Its ability to form a weak basic solution in water made soda ash ideal for washing the body and various fabrics as well as producing the earliest soaps.

In 1783, when the production of soda from ash could no longer fulfill French industry's growing demands, King Louis XVI offered a large cash reward for anyone who could find an efficient way to produce soda ash from salt, an inexpensive raw material that could be easily obtained from the sea. In 1791, Nicolas Leblanc was named the winner when he built a plant that succeeded in producing sodium carbonate from salt, sulfuric acid, carbon, and calcium carbonate. However, with the advent of the French Revolution, he never received the prize money, and his plant was confiscated by the revolutionaries. It was only when Napoleon became leader that the plant was returned to him, but Leblanc, unable to afford the repairs or compete with other soda ash manufacturers who had built their own plants, killed himself in 1806.

The Leblanc process was adopted in many countries around the world and used almost exclusively until 1863 when Belgian chemist Ernest Solvay invented a different process that is still in use today. The Solvay process is based on three readily available ingredients—salt, ammonia, and calcium carbonate—and has the advantage of costing less while not generating any contaminating byproducts like the irritating hydrogen chloride gas and malodorous solid waste produced by the Leblanc process. The first Solvay sodium carbonate plant was built in Belgium in 1864. In a matter of decades, every last Leblanc plant closed its doors, replaced by Solvay production. Three quarters of the soda ash currently used worldwide is the product of this process, while the remaining twenty-five percent is extracted from mineral deposits.

From carbonate to bicarbonate

The Solvay process is also used to manufacture sodium bicarbonate, or baking soda. The baking soda is actually an intermediate product that is still contaminated with ammonia when the process is complete. Therefore, to eliminate remaining ammonia residue and fully transform it into carbonate, it's heated. Then, it's treated with carbon dioxide to turn it into pure sodium bicarbonate.

3

Limescale

I own a lot of mugs. Many are souvenirs from trips I've taken, and I use them in rotation for my morning tea and the selection of aromatic herbal infusions I enjoy in the afternoon and evening. Two of these mugs, the oldest and most used, are white on the outside with patterns that have faded from the hundreds of trips they've made in the dishwasher. Inside, however, they have a brownish film that consistently resists these trips and will come off only with some heavy elbow grease and a good scrub. That, or a little chemical magic.

That said, I left these two old mugs the way they were and continued to use them for my daily infusions because I kept promising myself I'd make a video explaining what the unpleasant brown staining is. The video still hasn't been made. I'll get around to it someday, but in the meantime, for those of you who have experienced something similar, let me reveal the culprit: limescale. Calcium carbonate, possibly with a little magnesium carbonate, too—which, infusion after infusion, mug after mug of tea, clings to the inside. Limescale is white, but the vegetable pigments in herbal tea give it a brown color.

If the water supplied to your home is rich in calcium and magnesium salts, you'll no doubt have noticed other limescale deposits around your house: When dry, the glass shower door has white marks on it or tiny grains sticking to the surface. Glassware has a thin layer of white film that won't budge with soap and water. The toilet bowl has darker streaks, and there's a white crusty layer around the sink's metal faucets, not to mention the faucet's aerator. If you use the bathtub more than the shower, you'll notice some unsightly and difficult-to-clean darker streaks around the water line. In the kitchen, there might be a gray-white powder crusting up your pasta pot, kettle, and coffee machine. Limescale also clings to places you can't see: throughout the pipes bringing water to and from your house, inside the dishwasher and washing machine, on the showerhead (the holes of which might be completely stopped up), and so on.

Before I go on, I want to make it clear that absolutely no harm will come to us from drinking water with calcium and magnesium in it. Limescale might look unsightly on the sink and cause problems with your home appliances, but it is completely harmless to humans. Quite the opposite, in fact: Our bodies require magnesium and calcium salts as part of our daily recommended intake. We're not household appliances! People used to think that drinking the limescale in our water caused kidney stones, but now we know that this is not true.[1] Calcium oxalate is the substance causing all the problems, not calcium carbonate. Crazily, I often see

1 There are several studies backing this up, including Bradley F. Schwartz et al., "Calcium nephrolithiasis: effect of water hardness on urinary electrolytes," *Urology* 60, no. 1 (July 2002): 23-7.

people buy mineral-free bottled water while also spending money on calcium supplements.

Moving on, let's think about what we can do to remove limescale. Essentially, there are two options: Either we stop water with a high mineral content from coming out of the faucet or showerhead, or we eliminate the limescale deposits as often as we can. Let's start with the latter and talk about removing deposits from appliances like the washing machine later.

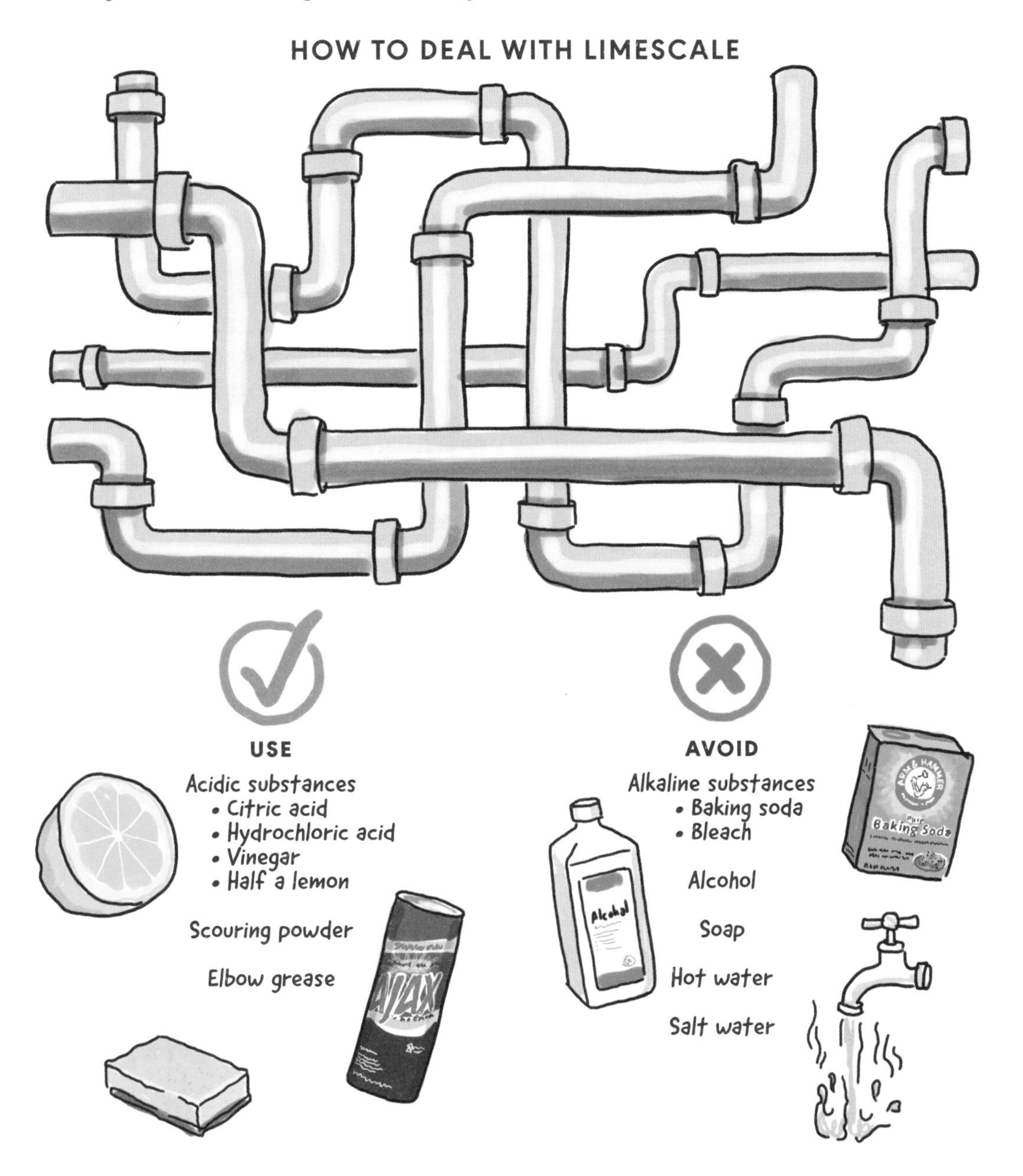

THE CULPRIT (AND ITS CHEMICAL SOLUTION)

Chemically speaking, limescale is primarily calcium carbonate. All carbonates react with acidic particles in water, dissolving and producing carbon dioxide.[2] I bet you have baking soda somewhere at home, most likely in the kitchen. When you combine it with vinegar or lemon juice (both acids), it reacts, releasing carbon dioxide and producing a different salt: sodium acetate or sodium citrate respectively.

My grandmother Lucia used to keep an orange- or lemon-flavored white granular powder in the kitchen cupboard. (That same formulation is still sold today as an antacid for relieving heartburn and indigestion.) I'd sneak into the kitchen to put some on my tongue, firstly because I liked the taste and secondly because I loved how it fizzed, which I now know was due to carbon dioxide released by the baking soda in the granules. My grandmother's secret potion contained sugar and an acidic substance that reacted with the carbonate when it came into contact with my saliva. If I'd swallowed it, the acidity of my stomach would've made the bicarbonate react and release the carbon dioxide in a different way: as a burp!

I'm sure you've got some milk of magnesia somewhere at home, too. Milk of magnesia is another antacid made from compounds containing magnesium—either magnesium hydroxide or magnesium carbonate.

Similar to how baking soda reacts when it comes into contact with an acid by releasing carbon dioxide, limescale does the same. But which acid? Delimers for home or industrial use contain several different types, and it doesn't really matter provided the concentration is strong enough to lower the pH and dissolve the carbonates.

My grandmother's trusted remedy was vinegar: easy to find at home, not too expensive, and not too corrosive (we drink it, after all). It works wonders on small deposits and easily removes limescale marks on windows. However, it's not as successful at removing thicker layers built up over more time because vinegar is only a weak acid solution. Being acidic, it dissolves limescale and produces carbon dioxide, but it can't prevent the limescale from reforming. And it smells! Splashing it on a salad is fine, but I don't like using vinegar to clean because the scent gets everywhere.

There are specialist products that eliminate limescale—from sinks, toilet bowls, and home appliances—and then capture any future limescale deposits before they settle on a surface, most frequently the inside of your washing machine.

---

2 Technically, carbonates don't actually dissolve when they meet acids; they react and transform. That said, this book attempts to use everyday language without unnecessary technical jargon.

WHAT'S THAT SUBSTANCE?

CALCIUM CARBONATE

Calcium carbonate, which has the chemical formula $CaCO_3$, is everywhere. Many of the stones used to build houses contain it—they're called limestone because they're composed of the most common crystal forms of calcium carbonate: calcite and aragonite. Marble falls into the limestone category, too—maybe your feet are resting on some right now. Travertine is limestone, while dolomite contains both calcium carbonate and magnesium carbonate.

Calcium carbonate comes out of the kitchen faucet, and it's an essential part of our bones. Don't go thinking we only absorb it from water, either: Spinach and other dark leafy green vegetables contain a good quantity of calcium. Calcium carbonate is in your fridge as well: Milk and milk derivatives are another good source because calcium (in the form of calcium phosphate) is the glue that holds their casein (the fundamental milk protein) together. Eggshells consist primarily of calcium carbonate crystals and a small amount of protein. The same goes for the shells of mussels, clams, and other shellfish. If you own a pearl necklace, you have calcium carbonate there, too.

You might have calcium carbonate-based antacid tablets in your medicine cabinet, or maybe you've sprinkled calcium carbonate on your lawn to make the soil less acidic. Of all the chemical transformations we might come across, calcium carbonate reacting with acids is the most common and the most widely used. In this reaction, calcium carbonate neutralizes the acid and ceases to exist, leaving behind calcium ions (Ca^{2+}) and releasing water and carbon dioxide.[3] If you have a swimming pool, one product you likely add regularly to keep the water clean and healthy is calcium carbonate, which stabilizes the pH and offsets the acidity of some disinfectants. Industrially, calcium carbonate has multiple uses. Not only is it a key raw material in construction, it's also the number-one ingredient in the production of glossy paper and paints, and it is even added to an assortment of plastics to change their mechanical properties.

Calcium carbonate powder is a mild abrasive, so it's often added to toothpaste because it doesn't damage our much harder tooth enamel. Lastly, as a food additive, it is designated as E170. E100 and E199 are normally employed as colorants and, likewise, E170 is used for its white hue. More often than not, it acts as an acidity regulator (for example, in canned vegetables). Calcium carbonate is also added as a source of calcium to vegetable milk substitutes made from soy, rice, oats, and so on to mimic the calcium content that would normally be present in dairy milk.

3 For chemistry students: The reaction is $CaCo_3 + 2H^+ \rightarrow Ca^{2+} + CO_2 + H_2O$.

PRODUCT SPOTLIGHT: **VIAKAL**

Viakal is a brand from Procter & Gamble. Whatever its form (spray or liquid, for bathrooms or stainless steel), it always contains water and two main ingredients: citric acid and formic acid. If we take a look at the ingredients of the classic Viakal liquid limescale remover, the official safety data sheets disclose that both acids have a concentration of between 1 and 5 percent. The solution has a pH of 2.2, making it more acidic than household vinegar.

Citric acid occurs naturally in a variety of fruits and plants and is used as an acidifier and pH regulator in a range of processed foods, not to mention in lemon and other citrus juices. Formic acid is produced by ants and was isolated for the first time in 1670 by distilling a large number of the insects. Not to worry, though—formic acid is now manufactured industrially through chemical synthesis in the same way that citric acid is clearly not extracted from individual lemons but rather mass-produced through bacterial fermentation.

Formic acid is corrosive to our skin and other organic matter, which is probably the reason it's combined with citric acid in Viakal's formulation: It helps remove any organic matter trapped in the limescale. These two acids are the main ingredients in this product, and without them, it would be unable to do its job.

After the acids in the list of ingredients come the corresponding sodium salts: sodium formate and sodium citrate. Next is sodium hydroxide (lye)—likely added to obtain the required pH—which reacts with the two acids to form the two salts.

The citrate ion remains when citric acid reacts with the limescale; as it's negatively charged, it captures the calcium ions and holds them in solution so that they can be washed away when the surface is rinsed.[4] This is why almost all descalers contain citric acid. (The acetate ion that forms when vinegar is used to remove limescale is not as effective at getting rid of lime.)

This brings us to the surfactants. Their role is to help dislodge the soap scum that builds up alongside limescale. A number of different surfactants are used, and these have changed frequently over time. The current formulation features ionic and nonionic surfactants, some of which may have been added to make a glossy film after the limescale is removed. If water can run off your shower glass more easily, it will be harder for limescale to redeposit.

If we were just removing limescale from the sink, we could probably stop here, but for limey streaks down vertical surfaces, a water solution alone isn't enough because it slides off too quickly. This is why substances like xanthan gum are added to give products like Viakal the viscosity they need to adhere to surfaces long enough to clean them.

Next up we have the various natural or synthetic scents that give Viakal its characteristic fragrance, followed by colorants. All of these are

4 We chemists say that the citrate "complexes" the calcium ions.

crucial because the blue color and strong smell make sure we don't distractedly pour some into a glass, forget we've done it, then take a drink.

As you can see in the table below, each ingredient has its own special function. But you won't find any of this written on the label—my descriptions are merely my informed interpretations of what the manufacturer most likely meant them to achieve, and they may also have been selected for other reasons.

Can I make it at home?

Sort of. To reproduce the formulation in its entirety would be difficult, though: Even if we left out a few ingredients, like colorants and fragrances, it would still be hard to get the right dosage of surfactants and thickening agents. Frankly speaking, I'm not sure it would be worth the bother. It's much easier to buy something designed especially for the job by chemists.

That said, acids are the main ingredients, and if you'd rather not buy a specific cleaning product, then in most cases limescale can be removed fairly easily with a homemade remedy of the kind our great-grandparents might have used. Indeed, most of us have at least one acid, albeit a weak one, at home. Who doesn't have vinegar? Acetic acid and other aromatic substances left over from fermentation give this household staple an acidity level of 6 to 8 percent. And my own favorite cleaning acid, citric acid, is found in lemons and oranges. So, before you throw away that lemon half you've just squeezed, you can always rub it on the limescale around your sink.

It's much easier, though, to buy a jar of citric acid—available online or in large, well-stocked grocery stores as an odorless white powder—and dissolve some in water to get a solution with the required concentration. No, it won't smell like lemons. But it's a non-volatile solid that doesn't release pungent odors, and since you're free to mix it in whatever ratio you like, you can create solutions that are stronger than vinegar. The resulting product can be stored in a plastic bottle. Depending on what I need to

WHAT'S IN VIAKAL?[5]

INGREDIENT	FUNCTION
Water	Solvent
Citric and formic acids	Acids that dissolve limescale
Undeceth-10, deceth-n, and sodium xylene sulfonate	Surfactants that increase the solubility of poorly water-soluble substances
Sodium citrate, monosodium citrate, and sodium formate	Salts present in the acidic ingredients to regulate pH; citrates can complex calcium ions
Sodium hydroxide	Regulates the pH to the required value and forms salts in the original solution
Xanthan gum and modified starch	Thickening agents that increase the solution's viscosity to make it stick to surfaces
Perfumes (hexyl cinnamaldehyde, benzyl salicylate, and limonene)	Gives the product its characteristic fragrance
Colorant	Gives the product its blue color

5 From info-pg.com and pgregdoc.com.

clean, I make up 5 to 15 percent-strength solutions (which means they contain between 50 and 150 grams of acid in one liter of water).

Can I buy a vinegar-based product instead?

"My grandmother swore by vinegar for removing limescale, so I'd much rather buy a natural vinegar-based product than a bunch of complex chemicals with names I can't pronounce." This is a popular comment that stems from a very understandable reaction to "chemophobia," aka fear or suspicion of anything with a chemical name (and by "chemical name" I mean the kind of incomprehensible tongue twisters we chemists use to identify molecules).[6] There's no rational basis for this fear, but it's fairly widespread nevertheless—a natural human reaction to things we don't know or don't understand. It also explains why manufacturers would never dare put "contains cinnamaldehyde and sodium xylene sulfonate" in bold letters across the front of the packaging. Instead, they generally list—in the tiniest print—only the ingredients they are legally required to disclose.

There are a number of chemical substances that consumers do not seem to see as actual chemicals. It's as if they're not molecules like all the others—instead, they seem to belong in a different category. To a chemist's ears, none of this makes any sense: All molecules are chemicals, and their origin (whether synthetic or natural) does not alter the negative or positive effect they can have on us or on the environment. Yet I often hear or read comments like, "I prefer to use vinegar instead of chemicals," as if vinegar wasn't a water-based solution of acetic acid. Vinegar is something we're familiar with, and it's been with us for thousands of years, so we think we know it well and don't fear it. Yet that quickly changes when we simply use its more formal, less familiar chemical name: CH_3COOH. Our suspicions grow even worse with its official chemical name: ethanoic acid.

Marketing and advertising harness this phenomenon to their advantage. How? The next time you go to the grocery store, check how many products advertise the fact they contain vinegar, baking soda, or worse still, both vinegar *and* baking soda (which, as we saw in the previous chapter, actually cancel each other out chemically). You might also notice that nearly all of them don't make any claims regarding these substances' cleaning power or properties—they simply state that the substances are present.

Consumers who are more susceptible to this kind of subliminal planting of information will connect the dots and draw their own conclusions.

Limescale removers in particular often have "contains vinegar" emblazoned across the front of the packaging.

> ⚠ **WARNING!** When mixing and storing cleaning solutions, I always recommend clearly labeling the bottle with its contents and preparation date. Also, it's a good idea not to use the same bottles you'd normally drink from because weeks or months later, you could easily forget that they don't actually contain milk or water. I recycle empty cleaning product containers, remove the original label, and stick on my own.

6 Ruggero Rollini, Luigi Falciola, and Sara Tortorella, "Chemophobia: A systematic review," *Tetrahedron* 113 (May 2022): 132758.

However, if you look more closely at the list of ingredients on the manufacturer's website, the descaling power always comes primarily from citric and formic acids. Number two on the list are the surfactants, followed by the acetic acid, which is only listed on the back so it can be shouted about on the front. In other words, the product is still the same thing, just with an extra ingredient to attract customers' attention. The manufacturers are not doing anything untoward—it's a perfectly legitimate marketing ploy to leverage individual preferences (and fears) for the purpose of commercial appeal. The same thing happens with shampoo: A standard formulation might feature jasmine tea in one variety, orange blossom in another, and oats harvested by virgins under a full moon in a third—all iterations of exactly the same shampoo.

DESCALING ACIDS

Chemists normally split acids into two categories: weak and strong. Strong acids, particularly in high concentrations, are extremely corrosive and should be handled with the utmost care. Always wear gloves and safety glasses to protect yourself when working with them. Since they're so aggressive, they are rarely used at home, but they are generally available in grocery or hardware stores and are often an ingredient in commercial cleaning products.

The most common strong acid is hydrochloric acid, also known commercially as muriatic acid, which is sometimes added to toilet cleaners and descalers for particularly stubborn deposits around toilet rims. Watch out, though, as similar-looking products can contain either an acid like hydrochloric acid or a base like bleach, with the only distinguishable difference being the color of the packaging—so always check the list of ingredients before buying.

STRONG ACIDS	WEAK ACIDS
Hydrochloric (muriatic) acid	Hydrofluoric acid
Sulfuric acid	Acetic acid (vinegar)
	Organic acids

Another common acid is sulfuric acid, which is highly dangerous—I urge you not to use it in your home. It's found in products sold to unblock sinks because of how powerfully it reacts with organic material. Chemical labs also handle strong acids like nitric or perchloric acid, but there's no need to use them for household cleaning.

Now, let's turn our attention to weak acids—but don't be fooled by their name, since "weak" doesn't necessarily mean "safe." Hydrofluoric acid is considered a weak acid for chemical purposes, but it corrodes glass and can be hazardous if it touches your skin. Thankfully, the weak acids we tend to encounter in our homes are all relatively harmless, and some we even eat or sprinkle on our food as a seasoning.

Vinegar is a weak solution of acetic acid in water that contains a different mixture of aromatic substances depending on how it's made. Lemon juice contains citric acid, yogurt contains lactic acid, and fruit contains citric, malic, succinic, and tartaric acids, to name just a few. Cleaning products are often made with these organic acids, or with formic acid. Weak acids in a water solution work more slowly than hydrochloric acid, but they can remove limescale deposits from surfaces that would be ruined by stronger acids.

GETTING RID OF LIMESCALE AROUND THE HOUSE

Coffee makers

If, like me, you have a coffee maker at home like me, no doubt you'll have seen limescale in the water tank. If you live in an area with fairly hard water ("hard" means "high mineral content"), you may have noticed limescale in your pipes as well. This buildup slows down the flow of water until the pipes become completely clogged. Luckily, limescale in your coffee maker is an easier problem to address.

To get limescale out of the inside of your coffee machine, the best place to start is with the instruction manual. I realize this may not be everyone's favorite thing, but I actually find instruction sheets quite helpful—there's often so much useful information to be gleaned from their pages. I routinely read the user guide before I get anywhere near switching on a new appliance.

Take my espresso machine, for example. Reading the descaling instructions, I learned that the manufacturer advises against using vinegar to clean the inside. This must be due to the risk of damaging the inner workings, like seals or metal parts. I was told to purchase a special descaling kit instead. While inspecting the contents of these kits, I spotted lactic acid and a small amount of detergent among the ingredients. Users are directed to add the kit to the water tank and run the special descaling procedure. On completion, the tank should be refilled with clean water and the cleaning mode should run again to flush out any leftover descaling solution.

Other manufacturers offer a variety of organic acid solutions to descale their coffee machines. These include malic, sulfamic, and citric acids. Not one manufacturer, as far as I am aware, recommends vinegar. If you'd rather avoid the minimal cost of purchasing the manufacturer's recommended descaling kit and decide to go ahead with a homemade remedy, then try this at your own risk: Fill the tank with a 5-percent solution of citric acid in water (5 grams of citric acid for every 100 milliliters of water) and leave it there overnight.

Online, you'll see a range of different citric acid solutions discussed, going all the way up to 18 percent. There's no reason for such a high concentration of acid, especially considering the risk of damaging your coffee maker's internal parts. It's a much safer idea to use a weaker concentration and repeat the process several times using the same solution.

To do so, pour your homemade descaling solution into the water tank, remove any trays or parts used for holding cups, and place a jug underneath the coffee outlet that's large enough for the full contents of the water tank. Run the machine until the tank empties. Fill the tank again with more descaling solution and repeat. Then rinse the tank, refill it with fresh water, and run it once more until empty. Don't throw the used acidic solution away—it can be reused for further cleaning. I wouldn't put it back in the original bottle, though, so maybe have a look around the house for anything else that needs descaling. Your sink's faucets, perhaps?

DID YOU KNOW?

Fizzy drinks like Coca-Cola and Pepsi are acidic. This is due to the phosphoric acid (a weak acid) added to offset sweetness. These drinks' pH is usually between 2.5 and 3, which is low enough to affect limescale. However, before you start squirting fizzy drinks at your sink, remember that these beverages contain many other ingredients, so you could end up with a lot more mess to worry about. What's more, their pH is higher than ordinary lemon juice or a specially prepared citric acid solution, so they are ultimately less effective.

Now that your machine is all sparkly clean, why not try low TDS (total dissolved solids) water, also known as low mineral water, for your next coffee? I buy bottled water with total dissolved solids of less than 50 milligrams per liter. Prevention is better than cure!

Faucets

What makes faucets so special that they get their own section? Nothing really, as we're still dealing with the same limescale as before. It generally deposits around the spout but can also form along the metal body. If you're using a commercial cleaning product, the thickening agents will make sure the solution adheres long enough to be effective. If you'd rather go for a homemade remedy, you can spray a little citric acid onto the faucets, but most of it will run off. This is not helpful, as the solution needs to be in contact with the limescale for at least five to ten minutes (depending on the concentration) to have any effect. Something that works well is soaking a paper towel in the acidic solution and wrapping it around the faucet, giving the citric acid the time it needs to get to work on dissolving the chalky crust. Simply remove the paper towel and rinse off the faucets afterward. You may find they still look grimy, but this is just dirt that was previously trapped in the limescale and unaffected by the acid. This is where the classic combination of an abrasive sponge, detergent, and good old elbow grease comes into play. Never underestimate the force[7] . . . of elbow grease!

> **DID YOU KNOW?**
>
> The speed and effectiveness of an acid-based limescale remover depends on the pH and concentration of the acid and the temperature you use it at. Generally speaking, the colder the environment or solution, the slower the limescale will dissolve, and the more patient you'll need to be.

Showerheads, walls, and doors

The shower is yet another spot limescale likes to burrow into—especially two places in particular: the showerhead and the walls, whether they're tile, glass, or plastic.

If the holes in your showerhead are somewhat clogged and water is spurting out in strange directions, then it's time to remove the limescale outside and the hard clumps inside. Unscrew the showerhead and start by shaking out any grainy residue or lumps. Before you treat the showerhead, check to see what the instruction manual says about acids and potential damage to chrome-plated or painted parts. Given its odd shape, it would take quite a large bucket of solution (whether store-bought or homemade) to fully immerse a showerhead. I find it much easier to pour a little liquid into a plastic freezer bag, place the showerhead inside, seal the bag around the neck with a rubber band, and set it face down so the holes that need to be descaled are all fully covered by the solution.

What about shower walls and doors? If you're dealing with tile, especially if there are also traces of mold (which is often the case), then my advice is to purchase a specific cleaning solution. Some products even claim to prevent limescale and mold from reforming after removal by leaving a film that carries water safely away. Spray clear doors, whether plastic or glass, with a descaler before rinsing them. But always read the manual first, as some glass doors have a special anti-fog film (just like some mirrors) and might not withstand the acidity of a standard descaler. In this case, your only option is a glass cleaner and elbow grease. If alkaline detergents are also not recommended for the glass, then try water and a little dishwashing liquid. You have limited options when it comes to removing limescale effectively: either acid or a good scrub. If neither of these work, try a damp microfiber cloth sprinkled with an abrasive

7 Said by Darth Vader in *Star Wars: Episode IV - A New Hope*, directed by George Lucas (Lucasfilm Ltd., 1977).

cleaner. Be careful, though—depending on how hard the glass is, you could permanently scratch it. The same goes for plastic doors. I suggest you test a small area at the bottom corner first.

Ironically, the abrasive cleaners sold to clean and polish surfaces often themselves consist of calcium or magnesium carbonate (the same solids that form limescale). Chalk powder is one of the softest materials that we chemists classify as inorganic, and it does not scratch glass. Starch is only slightly harder, and baking soda is similar to chalk in this respect. Of course, grocery stores offer a multitude of options for this type of cleaning—read the labels carefully, though, to check the list of ingredients and the types of actions they perform (whether chemical, mechanical, or both).

Some surfaces (glass or plastic) may be so dull or mottled that they are past cleaning, just like drinking glasses can corrode from one wash too many. The clouding on the glass is not a deposit or film on the surface—the surface has become permanently opaque because the original shiny layer has been etched away. If this has happened, then it's too late.

Bathtubs

Hard water doesn't just leave a layer of calcium carbonate in your bathtub—mixed with soap and most detergents, it also forms insoluble salts that "precipitate" (stick around!) and form nasty gunk. If you regularly bathe in hard water, you may have already seen an unsightly and very stubborn gray watermark lining your bathtub. Soap scum, as this chalky buildup is often referred to, collects on the surface, often trapping dirt (which can include sebum and dead skin cells). How to get rid of it? There's no easy solution, unfortunately. Soap scum is insoluble, so only serious scrubbing with an abrasive detergent will do the trick, maybe after attacking it with a 10-percent citric acid solution. Be careful not to ruin the surface of your bathtub, though.

WHY DOES HOT WATER CAUSE LIMESCALE?

You may have also noticed a chalky buildup inside your favorite saucepan, usually around the water line. That's limescale, too. Why does heating water, especially in areas where the water is particularly hard, cause limescale to form almost instantly?

Minerals are generally more soluble at high temperatures. Take cooking salt, aka sodium chloride, for example: Slightly more dissolves in hot water than in cold. For some substances, like sugar, the difference is enormous: Almost four times the amount stirs nicely into hot rather than cold water. So why would calcium carbonate be any different? The carbon dioxide in these substances loves to dissolve in water. Unsurprisingly, the fizz in popular fizzy drinks comes from CO_2 as well rather than nitrogen or any other gas.[8] The opposite is normally true for gases: Solubility drops as temperature rises. When water is heated, the dissolved carbon dioxide escapes. At the same time, the high temperature turns bicarbonates into carbonates, causing further release of carbon dioxide.[9] (You can learn more about where these bicarbonates come from starting on page 46.)

Calcium bicarbonate only exists as a solution—it cannot be isolated. Think about it like this: As long as calcium ions in water are surrounded by bicarbonate ions, both remain dissolved. If the bicarbonate ions were to disappear and be replaced with carbonate ions, then the latter would immediately bond with the calcium ions to form calcium carbonate (CaCO3), which is poorly soluble. As chemists would explain it, this would result in the calcium carbonate's precipitating from the solution to form a solid residue. This is precisely what happens when water is heated.

If your water is rich in bicarbonates, these can be removed by simply boiling the water and pouring it into a new container. This lowers the water's hardness, but

DID YOU KNOW?

Common soap does not lather easily in very hard water. This is because the calcium and magnesium ions in the water (positively charged) react badly with the surfactants in the soap (negatively charged) and stop the soap from doing its job. For the opposite reason, rinsing soap off your hands in overly soft water is equally difficult!

8 Some beverages, like Guinness, use nitrogen and carbon dioxide.

9 $2HCO_3^- \leftrightharpoons CO_3^{2-} + CO_2 + H_2O$.

only in terms of bicarbonates. (Hardness due to the presence of bicarbonates is generally referred to as "temporary water hardness.") Heating has no effect on other calcium salts like chlorides and sulfates, which are left behind when water evaporates.

Collateral damage

Acids eliminate limescale, but beware: They can corrode other materials at the same time, especially those with low pH and high acidity. Metals are subject to corrosion from concentrated acids, and I'm guessing you definitely don't want to ruin your nice stainless steel when removing limescale residue. Glass and ceramic are much more resistant to acids (although not all of them), so you can worry a little less about these materials. However, carbonate stones like marble, travertine, tuff, and dolomite should absolutely never be cleaned with acids. The limescale would slide off them for sure, but it would take a good part of the original surface material with it. For chalky deposits on this sort of surface, the safest method is buying a specialist product (either neutral or mildly basic) and arming yourself with a bucketload of patience and, as usual, a healthy dose of elbow grease.

Bottled water

What type should you buy? I'd suggest a low fixed residue (less than 50 milligrams per liter) water for coffee machines or similar appliances with water tanks that are subject to limescale buildup. For thirst-quenching, it depends on your personal preference, but maybe give tap water a go first. There are no health benefits to drinking low fixed residue water purchased at the store.

The taste of hard water

I was taught in primary school that water is odorless, colorless, and flavorless. In terms of odor, I'm pretty much in agreement: Any smell I might detect no doubt comes from dissolved volatile compounds that have escaped and floated over to my nose. Likewise, I have few objections about color as long as we're talking about small amounts confined to a glass. It can get a bit more complicated with larger quantities, like seas and oceans, which can appear blue, but that's a story I'll save for some other book (and no, it's not because they reflect the blue sky). But would any of us be willing to claim, hand on heart, that water has no flavor? If you live in an area with very hard water, you bet there's a flavor, just like you can taste it when water is rich in iron. Maybe distilled, pure water has no taste, but that's not the kind of water we actually drink. Regular drinking water has a different flavor depending on the quantity of salts dissolved in it—and calcium and magnesium ions have a very distinguishable taste.

HARD WATER

We often hear people talk about "hard" water, but what does this actually mean? Hard water is simply water containing a high percentage of minerals (as compared to soft water, which has a low percentage). Rainwater is often very soft because of the scarcity of mineral content, and so are deionized and distilled water available at the store. We're used to seeing soft water as the opposite of salt water. But since we don't have seawater coming out of our faucets, in household terms, sodium chloride is of less interest to us than other types of salts, namely calcium and, to some degree, magnesium. We'll also learn that non-salty water—despite being commonly referred to as soft water—can also be hard.

There are two components to water hardness: permanent hardness and temporary hardness. Permanent hardness is caused primarily by the presence of calcium and magnesium ions: the ones designated Ca^{2+} and Mg^{2+} on bottled water labels. To be more precise, permanent hardness also depends on the concentration of other ions with a positive charge greater than one, such as iron (Fe^{2+}) or aluminum (Al^{3+}) although, natural hot spring water aside, these and other metals can be overlooked for now since they're present in such small quantities. In water with high permanent hardness, calcium and magnesium ions are usually held in solution by chloride (Cl^{-}) or sulfate (SO_4^{2-}) ions. If you boil and then cool this kind of water, all the ions will still be there because the chloride and sulfate minerals remain in the pot. The only way to reduce or completely remove permanent hardness is via special equipment (domestic or industrial) or through chemical treatment, which is what happens in the purification plants that treat our public water supply.

Temporary hardness is caused by the same calcium and magnesium ions held in solution by hydrogen carbonate (HCO_3^{-}) and partially by carbonate (CO_3^{2-}) ions. In this case, boiling water with high temporary hardness destabilizes the hydrogen carbonate ions, producing carbon dioxide and releasing carbonate ions during evaporation. The remaining calcium ions are no longer soluble, and the calcium carbonate separates from the solution to leave a solid residue and soft water. It is this elimination of hardness that causes the buildup of limescale in your kettle each time you boil water.

HARD WATER	SOFT WATER
Most tap water	Rain and snow
Sparkling water	Distilled water
Well water	Deionized water

WHERE DOES THE LIMESCALE IN WATER COME FROM?

Chemistry is usually seen as quite tricky and is probably the second most-hated subject at school after mathematics. Chemical formulas and reactions are often learned by heart and recited like magical spells, and once school is over, they very quickly fade from our memory. But a single chemical formula tends to stick with us, and that's the one for water: H_2O, two hydrogen atoms combined with one oxygen atom. Most people are also aware that pure water is neither acidic nor basic but neutral, with a pH of 7.

This is only true, however, of water that has never been in contact with anything else. If you pour some pure water (for our purposes, let's take the distilled or deionized water on sale at the grocery store as pure) into a basin and leave it there for a few hours, when you come back, the pH will be slightly acidic at just below 7. What on earth happened in the meantime? It might seem strange because you can't see it occurring, but gases (not just solids like salt and sugar) also dissolve in water. This is how fish get the oxygen they need to survive. Oxygen isn't very soluble, though, and its solubility decreases as water's temperature rises. This is why adding hot water to an aquarium kills the fish inside by robbing them of the oxygen they need to keep breathing. On the other hand, as I mentioned earlier, carbon dioxide dissolves easily in water, which explains its widespread use in fizzy drinks.

Take a look at a bottle of sparkling water. You won't see any bubbles—not until you open the bottle and they all start to float up to the surface. What you're seeing is carbon dioxide leaving the water. The same thing happens when you open a bottle of champagne and so many bubbles burst out that the wine risks literally exploding outward. What causes these CO_2 bubbles to spring forth the minute you pull out the stopper or unscrew the cap? Pressure. In each of these cases, the pressure inside the bottle is greater than the atmospheric pressure outside. We probably don't notice this pressure differential when we open sparkling water, but we definitely do when we open champagne: The loud pop gives it away! Carbon dioxide becomes more soluble in water the higher the pressure (as do all gases), and in a bottle of champagne, the internal pressure can reach up to eight times atmospheric levels. This extreme pressure is what propels the cork out, and it requires the bottle to be made of thicker glass to prevent it from exploding.

So, to answer our question: When we open a bottle, the pressure inside suddenly drops to equal the outside pressure. The carbon dioxide, which was previously dissolved in water at high pressure, has to adapt to this new situation, and part of it is forced out of the solution, forming bubbles. However, the dissolved CO_2 has partly bound itself to water molecules to form carbonic acid,[10] which separates, like all acids, into hydrogen and hydrogen carbonate ions. These latter ions may release further hydrogen to form carbonate ions. Calcium carbonate is not particularly soluble in water, but calcium bicarbonate is.

But aren't we supposed to be talking about limescale? Why have I forced you to sit through a chemistry lesson on the solubility of carbon dioxide and the formation of hydrogen carbonate ions? Well, that's because these processes form the basis of those annoying white crusty layers than build up around your home.

> **DID YOU KNOW?**
>
> Dolomite is a sedimentary rock consisting primarily of calcium and magnesium carbonates. It is called "sedimentary" because it was formed hundreds of millions of years ago from the sediment left behind by ancient seas. This explains why marine fossils are often found imprinted in the rocks of the Italian Dolomites and many other mountain ranges around the world.

Water's journey

Let's follow a droplet of water that starts inside a cloud on a rainy day. Temperature and humidity cause the water vapor in clouds to condense, forming our droplet. It begins its descent to the ground as more or less pure water. On the way down, it encounters carbon dioxide that's naturally present in the air and dissolves some of that CO_2.

Then the droplet hits the ground and gradually seeps through the soil until it reaches a body of rock and/or sediment that holds groundwater. Not all rainwater travels this far, though. Some of it evaporates after the rain stops, while some feeds the surface water of streams, rivers, and lakes until they join the sea. But let's follow the droplet in its underground journey. On its travels, the droplet may come into contact with more carbon dioxide—including CO_2 released by organisms living in the soil—and dissolve more than it would be able to on the surface because of the increased pressure underground (just like the pressure inside a bottle of champagne). The droplet, turned acidic by the dissolved carbon dioxide, may encounter carbonate rocks (typically calcium and magnesium carbonate) at some point. As we saw earlier, calcium carbonate is not very soluble in pure water, but the droplet, now acidic, turns the material into a bicarbonate, which is much more soluble.

Multiple things could happen at this stage. While I'm tempted by my background as an explainer of all things chemical to tell you the story of how the droplet could become part of a breathtaking and gravity-defying

10 $CO_2 + H_2O \rightarrow H_2CO_3$.

stalactite or stalagmite in an underground cave, I have to keep reminding myself that this is a book about cleaning and chemistry in the home. So it's back to the droplet of water, now charged with calcium bicarbonate. The droplet heads to the public water system, which carries it at high pressure (set by the water company and always greater than the atmospheric pressure) straight into our homes and out of our faucets. As with the bottle of champagne, any excess carbon dioxide tends to be released at this point, while the calcium bicarbonate turns into calcium carbonate and limescale residue forms. Not all droplets travel though calcium carbonate deposits, which is why water's quality and properties can vary so much from place to place.

Removing limescale from water

If water contains too much dissolved calcium and magnesium, it can be treated before it gets to our taps, a process that water supply companies typically do for us. One very old but effective method is to introduce lime (calcium oxide). Performed for the first time in 1841 using water from the Thames, the process simply consisted of adding lime to water. Without going into the precise chemistry, lime makes water more basic by raising its pH to over 10.3, which precipitates the insoluble calcium carbonate. If necessary, the magnesium can also be precipitated as magnesium hydroxide by pushing the pH a little higher.

When a chemist states that a substance "precipitates" from a solution, it means that a solid residue forms at the bottom of the container. This often happens immediately, like with lime: The solid residue separates easily, leaving the water above it crystal clear. The increase in pH turns the dissolved carbon dioxide into bicarbonate ions first, then into carbonate ions which, in the presence of calcium ions, precipitate as calcium carbonate. A precise amount of lime must be added to guarantee the water is at the correct pH for drinking.

This process eliminates calcium and magnesium salts, which are present in the water as carbonates and bicarbonates. Other treatments, such as adding soda or sodium carbonate, are carried out to correct permanent hardness by removing large amounts of other types of calcium salts, such as chlorides and sulfates. The chemistry of water softening is a fascinating business, but I don't want to get carried away by my enthusiasm for the subject, so I hope this explanation will suffice for now.

The treatment described above lowers the amount of total dissolved solids in water, but it's not something that can be done at home. For domestic purposes, you can buy various ion exchange water filters, which need to be regularly refilled with salt crystals. These machines won't drastically reduce the overall amount of total dissolved salts in your water because the calcium and magnesium ions they remove are simply replaced with sodium ions. Other appliances, such as reverse osmosis systems, trap salts in a membrane while allowing filtered water to pass through.

4

Soaps

In chemistry terms, soap is the sodium or potassium salt of a fatty acid. Put like that, it doesn't seem like a chemical substance of any particular interest—it may even sound like nonsense to most. Yet the history of these compounds is wound up in thousands of years of culture and tradition from around the world. Keeping ourselves clean is a natural instinct that we share with most animal species: No doubt you've seen cats licking their fur, watched documentaries of monkeys taking turns cleaning each other, or been transfixed by tiny birds splashing in puddles and water fountains. Religions around the world have specific guidelines about how the body should be cleaned before particular ceremonies and occasions, not just for priests and other holders of important positions but also for congregation members. And since time immemorial, humankind has been coming up with new ways to bathe our bodies and clean our environments.

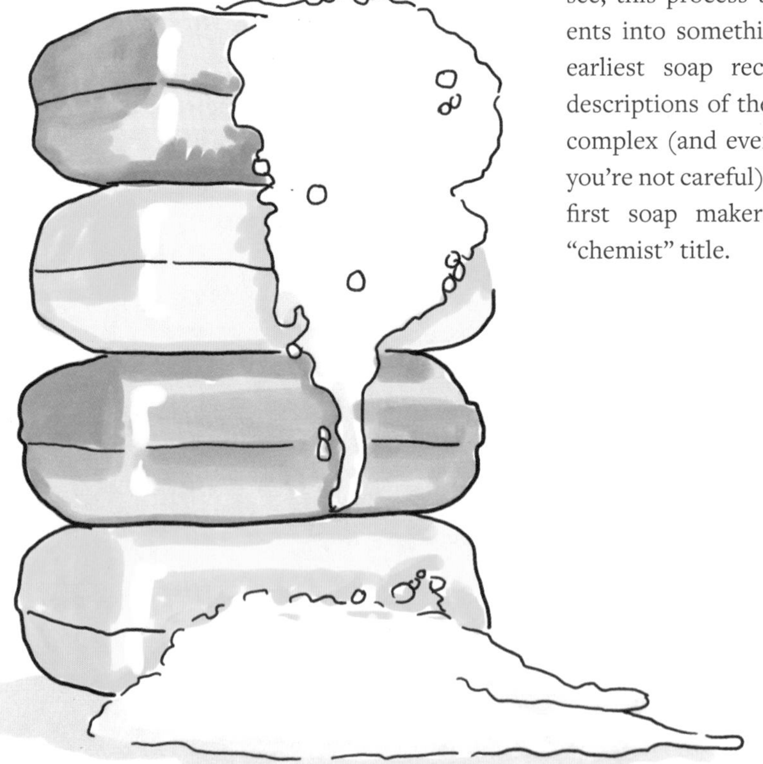

The invention of soap was a watershed moment in the evolution of cleaning technology. But a chemist like me also looks at the sodium and potassium salts of fatty acids and sees the progress my discipline has made, despite the fact that millennia passed before it was ever recognized as chemistry. In fact, soap was one of the first chemically synthesized products ever invented by humans. While advertisements may try to portray soap as "natural," it is actually the result of a chemical reaction called saponification that does not occur in nature. As we'll see, this process turns the original ingredients into something entirely different. The earliest soap recipes were basically just descriptions of the steps required to trigger complex (and even somewhat dangerous, if you're not careful) chemical reactions, so the first soap makers definitely deserve the "chemist" title.

EARLY SOAPS

The earliest written source describing a product used for cleaning that's recognizable as "soap" can be traced to a Sumerian tablet from 2200 BCE. The product was made from an alkaline substance combined with water and oil, and the resulting compound was used to clean wool. The Sumerians realized that if they wanted to dye fabric evenly, they had to find a way to remove the grease first.

Thousands of years ago, soaps were fairly primitive mixes of animal fat and vegetable ash. They may have been coarse, but they lathered and cleaned all the same. Early recipes were tried and tested over millennia and eventually brought us to products much more similar to the soap we recognize today. The chemistry of soap has remained largely unchanged in the last few centuries.

For hundreds of years, soap was made by adding fat to sodium and potassium carbonate from wood ash. The mixture was boiled for a long time to allow saponification to take place.

After the fall of the Roman Empire, public baths, running water, soap, and general personal hygiene became more of a rarity than a natural part of everyday life. Not surprisingly, some people refer to the Middle Ages as "the millennium with no bathroom." This lack of cleanliness in homes and public places, combined with poor personal hygiene, more than likely contributed to the spread of the many epidemics that struck Europe during this period.

To be fair, it's not completely accurate to say that soap disappeared from Europe during the Middle Ages: In some cities, soap production flourished, albeit on a smaller scale. Some of these places are still associated with soap today—like Marseilles, first and foremost, where soap of the same name is said to have been produced starting as early as the ninth century. A soap made in the Spanish region of Castile became famous slightly later. Generally speaking, all Mediterranean regions with ready access to olive oil produced soap to some extent. Savona and Venice did, and their processes were guarded religiously by various guilds and corporations. In this era, making quality soap from the available ingredients became more of an art than a science. It was only with the Renaissance and the social changes it brought that soap became more popular, especially among the upper classes.

DID YOU KNOW?

Soap flakes, invented in 1899, were the first home cleaning product not packaged in a cake or bar.

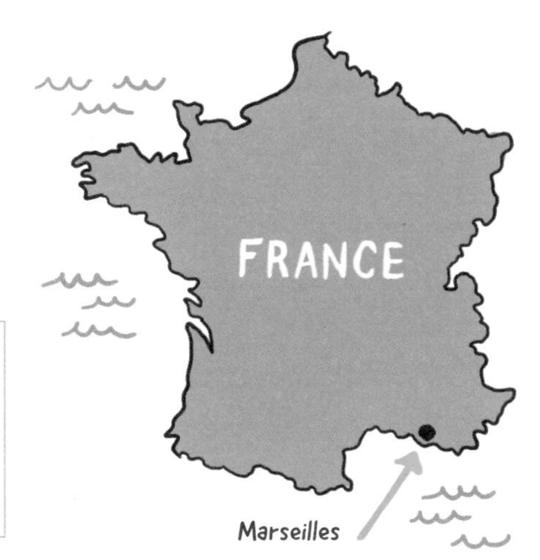

Saponification

Saponification is the chemical process of converting fat to soap. Fats can be either solid or liquid at room temperature. In the latter case, they're called oils—although, chemically speaking, oils are still a mix of triglycerides (three fatty acids bonded with a glycerol molecule). If you've ever been bored enough to read a household soap product's ingredient list, you may recognize some of the most common fatty acids: oleic, palmitic, and stearic. All of them must be freed from glycerol to produce soap, and this can be done in two ways.

In traditional soap production, the fats are mixed with a strong base, like lye or caustic potash,[1] and heated to make them react. The fatty acids break away from the glycerol, releasing it, and then the acids react chemically with the base to form salts (those sodium or potassium salts of fatty acids I've been referring to). On soap labels, you'll find names like sodium palmitate and sodium oleate, which are the sodium salts of palmitic and oleic acids separated from a triglyceride.

In the second method of soap production, the triglycerides are broken into glycerol and fatty acids. The latter are then further split up and made to react with a strong base.

BREAKING APART GLYCEROL AND FATTY ACIDS USING A BASE MAKES SOAP

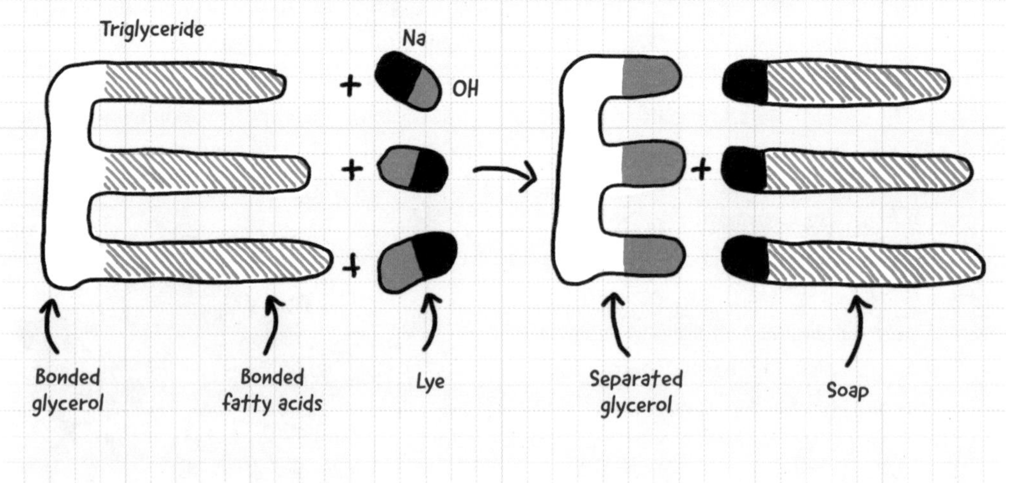

1 Caustic potash, aka potassium hydroxide, has the chemical formula KOH.

THE HISTORY OF LAUNDRY DAY

Nowadays, many people have washing machines at home and can do laundry whenever they choose. It isn't unusual, especially in larger families with several children, for clothes to be washed every day, accumulating piles of lights, darks, and colors, or even separate batches of cotton and synthetic fabrics so that full loads can be run each time. Considering all this, it might seem strange to think that, in the past, laundry was done much less frequently: once a month, or every few months in some rural locations. It was a laborious and extremely tiring job whether garments were washed in cold water from streams and rivers or boiled in large pots. The clothes were then dried in the sun (which often meant they were simply laid across the grass) for several days at a time to allow the ultraviolet rays to finish whitening any remaining stains.

Several days before laundry day, soap would have to be made if none was available. Ideally, this happened in the early spring, as fires and stoves had been burning wood all winter, creating ash that was stored in casks. Livestock were slaughtered in the winter, and the fat that ran off from the meat was collected and kept somewhere cool until it was time to use it. Ash was boiled in water for many hours, then filtered to collect the impure basic solution of sodium and potassium carbonate. Meanwhile, the fat was separated from the meaty tissue. This was done by heating chunks of meat over the fire and patiently waiting for the fat to drip off. This fat was then boiled with the alkaline solution for six to eight hours. To obtain a hard soap, salt was added at the end of the process, although more often than not this stage was skipped, since salt was an expensive and scarce commodity that was also required for a number of other things. Soap was left softer as a result, and then it was ready for use on laundry day. Ah, the good old days—not actually the proverbial bed of roses they're often made out to be.

FAMOUS SOAPS

A number of soaps are still associated with their city or town of origin despite no longer being produced there.

Aleppo

Without a doubt, one of the best known is Aleppo soap from Syria. Olive oil and up to 20 percent laurel oil are mixed to make this characteristic deep green block of soap with its unmistakable aroma. It is not known exactly how old the soapmaking tradition of Aleppo is, although according to one legend, Queen Cleopatra of Egypt practiced it, and it's been carried out for at least a thousand years.

Traditional production methods involve boiling the olive oil and the alkaline substance—sodium carbonate or hydroxide—for a long time, up to three days. While the newly formed soap is still in liquid form, laurel oil is stirred in and the mixture poured onto the floor, where it is left to cool before being flattened and cut. The resulting blocks are left to age for several months, during which time the soap hardens (by losing moisture), any remaining unreacted sodium hydroxide turns into carbonate, and the exterior turns gold.

Castile

Another Mediterranean location whose name is still associated with soap production is Castile in Spain. Soap from this region is also made from olive oil and comes in a hard block, but it's not green like Aleppo soap since it doesn't contain laurel oil. Like Aleppo soap, many stories are told about how Castile soap originated. One such legend says that eleventh-century crusaders stumbled across Aleppo soap and saw how it was made on one of their many military campaigns in the region. After returning home to Castile, they set about producing the soap for themselves (but without the laurel oil) and the tradition took root. Like Aleppo soap, Castile soap was also made by boiling olive oil with an alkaline substance, plus a brine added to separate the soap from the rest of the liquid.

We can never know how much truth there is in these legends, since quite often, producers are keen to give their products a history to increase their intrigue and appeal. What we do know is that people have been making soap since the end of the Roman Empire, if not earlier. But there are few traces of sites in Europe that produced or traded soap during the Middle Ages. For centuries, it remained a niche product, little used and rarely marketed, not prestigious or popular enough to make a city or region famous. Written

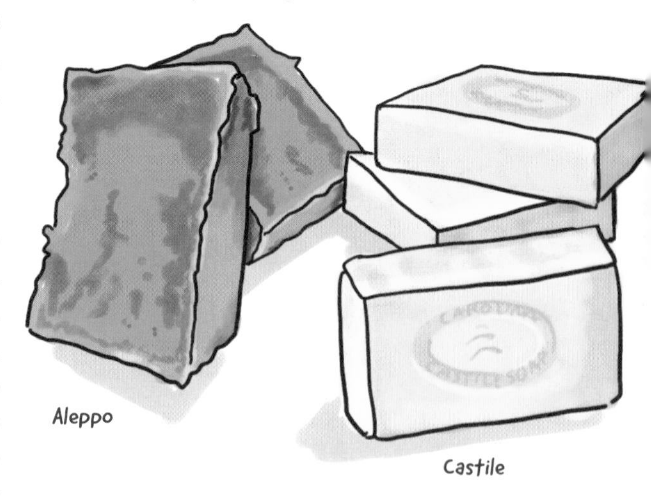

records document the production and export of Castile soap starting at the end of the sixteenth century.

Marseille

Marseille soap is similar to Castile soap. While there are traces of its production across the French region of Marseille during the Middle Ages, it took several centuries before it gained recognition and visibility throughout France. In 1688, Louis XIV, the Sun King, introduced regulations through the Edict of Colbert (Colbert was one of Louis' ministers) forbidding the use of animal fats in Marseille soap and requiring that it be made with 72 percent olive oil. Marseille soap was less aggressive than soaps made with other fats. This, combined with its solid square shape, made it a highly popular product. When large quantities of vegetable oils began to arrive in the port of Marseille from the French colonies in the nineteenth century (palm and coconut oil in particular), the olive oil in the original recipe was partly replaced. This resulted in a different product that was more off-white in color. In 1880, there were at least one hundred manufacturers of Marseille soap in France. Nowadays, there are far fewer. Small individual craft enterprises aside, there are only four large production sites with a workforce of around one hundred people and an output of only about 2,500 tons in 2015.[2]

Animal fat in soap

Animal fat is still used to make soap today. If you read the labels on the soaps sold in your local grocery store, you'll most likely spot at least one that contains sodium tallowate. Tallow is an animal fat, and sodium tallowate is the salt obtained when tallow reacts with lye. If you're vegan, make sure to check the ingredients before buying soap.

Marseille

Laundry soap

Laundry soap

Until the 1950s, in addition to the classic bar of soap for washing your body, the other omnipresent home cleaning product was the laundry soap bar. This product first appeared on the market in the mid-nineteenth century. If purchased as a block, it was often shredded later at home, as the flakes dissolved more easily in water. Laundry soap was an all-purpose product used for clothing, tablecloths, linens, floors, and anything else that needed cleaning. And if the smaller cakes of body soap ever ran out, despite their different formulation, laundry blocks would often be used in the bath as well!

2 Soap Professionals Union of Marseille, "Specifications for 'IG Savon de Marseille'" (December 2015).

Blocks of laundry soap were probably the most widely sold and used cleaning product in the world at one time—that is, until they were ousted by powdered detergents in wealthier countries. Laundry soap blocks' popularity declined rapidly due to the advantages synthetic products have over traditional soaps. Nevertheless, the classic laundry soap bar is still widely sold in developing countries, although washing powders are beginning to catch up as economies expand.

All that said, keeping a bar of old-style soap around the house is not a bad idea. It's like the workhorse of the cleaning world: simple and effective in an emergency. The high pH is probably not ideal for your skin, but it makes short work of a greasy stain on your best tablecloth that other liquid detergents struggle to fix. Just dampen the laundry soap and rub it over the stain on both sides of the tablecloth. Scrub gently, then rinse the cloth under warm water to wash away the soap (and hopefully the stain with it). If the stain persists, then it's probably not stubborn grease after all, but a colored substance, so read the sections on bleach in chapter 6 to find out what to do next.

THE ROAD TO TODAY'S SOAPS

Soap production did not evolve to any great extent while wood ash, and the lye obtained from it, was the primary alkaline ingredient. Soap as we know it today is the outcome of the mass production of soda ash (formally known as sodium carbonate).

Mass production

Nicolas Leblanc discovered how to manufacture soda ash industrially in 1791. His process was superseded in 1863 by Ernest Solvay, who found a way to produce soda ash more easily and more economically from sodium chloride and calcium carbonate through the use of ammonia. Around the same time, trade in fats—palm, coconut, and whale oil in particular—was gathering pace in international markets while cottonseed oil's star was also rising. During that same period (with a patent filed in 1908, to be precise), Procter & Gamble developed the fat hydrogenation process, which converts unsaturated fats into partially or fully saturated fats. This was key to how fats would then be used to produce soaps.

The final step that led to today's mass-produced soap occurred when two processes became distinct: removing glycerol from fatty acids and combining fatty acids with a strong base to make soap. An

DID YOU KNOW?

French chemist Michel Eugène Chevreul discovered the chemical composition of fats and oils, publishing his findings in 1823 and revolutionizing the science of soapmaking.

industrial-scale apparatus for the release of fatty acids from glycerol was patented by Procter & Gamble in 1939. This machine was called the hydrolyser, referring to hydrolysis, a technical process in which triglycerides are catalyzed in steel tanks containing water and fat. The fatty acids are first separated from the glycerol, and then the various types of acid—lauric, palmitic, stearic, and so on—are isolated. The underlying science was not new to chemists and engineers, but it took more than a century to develop the process on a large scale. Once the glycerol is removed, the conversion of fats into soaps is so fast that this new process cut soap-production time from two weeks down to less than a day.

From small candle makers to large multinationals

In the nineteenth century, candle and soap makers often worked side by side or were part of the same company, mainly because both products required tallow (the animal fat lining mammals' internal organs). Earlier, in the eighteenth century, candle production was more profitable, but soap manufacturing grew in popularity until it became the dominant activity, to the point that many of today's large multinational soap manufacturers were once just small candle makers.

In the early twentieth century, there were several hundred companies producing soaps. Many disappeared or were taken over by others, which grew to become the giants currently dominating the modern soap scene. No doubt you'll have heard their names: Colgate, Procter & Gamble, and Unilever.

William Colgate set up a small starch, soap, and candle business in New York City in 1806. Soaps at that time were rough, harsh on the skin, and fairly unpleasant-smelling. As decades went by, several small businesses merged into Colgate—including the B. J. Johnson Co., which launched a soap brand in 1898 that is still popular today: Palmolive. As its name suggests, Palmolive soap was made from palm oil and olive oil. This might come as a surprise due to the widespread misconception that palm oil was only recently introduced to the food manufacturing and cosmetics industries. The truth is, it has been a popular ingredient for more than a century. The success of Palmolive soap can also be traced to some very clever advertising

Producing lye

With the advent of the Leblanc process, which creates a much purer soda than that previously obtained from wood ash, it also became simpler to produce sodium hydroxide (also known as lye or caustic soda). Soda is made caustic with the addition of slaked lime (calcium hydroxide). Calcium carbonate precipitates to leave a lye solution that is very aggressive: A single splash of concentrated solution on your skin is extremely harmful.

suggesting that even Cleopatra used to wash with soap made from olive and palm oils.

Brothers-in-law William Procter, a candle maker, and James Gamble, a soap maker, founded Procter & Gamble (also known as P&G) in Cincinnati, Ohio, in 1837. Their business was an instant success. In 1859, they recorded sales in excess of one million dollars, and by 1878, they were producing twenty-four different types of soap.

Sixteen-year-old William Hesketh Lever started work at his father's Lancashire grocery business in 1867. In those days, soap was delivered from wholesalers in large brown blocks weighing a couple of pounds that grocers would cut up and sell to the public by weight. William decided to enter the soap market in 1884, founding Lever Brothers with his brother James. The industrial revolution had brought hundreds more people into the city and driven the emergence of a new class of workers—the so-called middle class—who were seen as more cultured and attentive to personal hygiene. All this suggested to the brothers that demand for soap was guaranteed to rise. Their first great success came in 1889 with Sunlight, a soap made using vegetable oils like palm oil. They also made a major break with tradition and stopped selling soap by weight, instead wrapping the bars individually and imprinting their brand name on the soap itself. In 1929, Lever Brothers merged with the Dutch company Margarine Unie to form Unilever.

DID YOU KNOW?

Scientists have only fully understood the complex physical structure of soap's liquid and solid phases for less than one hundred years. The materials we label "soap" are both solids and liquids, existing in various phases in which the salts of fatty acids are connected in different ways.

Advertising

The soap industry was one of the first to make heavy use of advertising to promote a packaged product with an easily identifiable name. In the late nineteenth century, most consumer goods—whether they were for cleaning the household or personal hygiene, or even food—were still sold loose. Soap manufacturers like Procter & Gamble focused their efforts on creating a recognizable brand by selling bars of soap wrapped individually with the product name clearly labeled. In 1904, Procter & Gamble spent $400,000 a year on advertising, which was an enormous amount in those days.

Not just for washing

What consumers want from soap hasn't really changed in the many thousands of years since its creation: They want it to clean, nothing more, nothing less. They want clean bodies and clean clothes. That said, in the late nineteenth century, soaps for bathing began to be charged with additional tasks, and new ingredients beyond the usual oils and bases were increasingly added to the formulations.

In 1872, Colgate introduced Cashmere Bouquet, the first-ever scented soap. P&G followed in 1879 with Ivory, a soap that could float on water—an iconic product that's still sold today. 1948 saw the first-ever deodorant soap, featuring the antibacterial substance hexachlorophene, which kills bacteria on skin and eliminates any unpleasant body odors caused by sweat.[3] From 1948 to 1967, there was a constant succession of deodorant soaps that looked after our skin. Then, the need to *feel* (not just *be*) fresh and clean emerged as a requirement between 1968 to 1993. Throughout the 1990s, soap was expected to hydrate, destroy bacteria, exfoliate, and soften, while also containing more and more natural ingredients. We're now in peak "eco era," with multiple "decoy" ingredients added for no other reason than to be trumpeted on the product label with bursts like "contains green tea" and "with avocado oil." Such ingredients usually have no actual purpose, and their quantities are generally not significant enough to make a difference.

TYPES OF SOAP

Historically, soaps have always been classified by their consistency: either hard or soft. This property comes from the type of alkaline substance used to make them. During the early Roman Empire, Pliny the Elder wrote about how the Germanic tribes made both hard soap ("sapo durus") and soft soap ("sapo mollis"). In terms of the underlying chemistry, a sodium-based alkali (carbonate or hydroxide) produces a hard soap, while potassium alkalis (carbonate or hydroxide) create softer soaps.

Soft potassium soap is not very commonly made these days, although it has experienced a revival of sorts on social media like Instagram and TikTok. It's pale yellow, transparent, and feels a bit like a firm jelly. And just like jelly, it cuts easily with a knife and can be quickly converted to liquid soap by adding water. Made with potash and olive oil, it dissolves easily in alcohol, and in the past, it was even used as an antacid.

Another way to classify soaps is based on how they are made: via a hot or cold process.

Cold processes harness the heat released internally when lye reacts with oils to produce a soap that is ready to use in a matter of days. What's more, the glycerol isolated from

HARD SOAP	SOFT SOAP
Made using sodium salts of fatty acids	Made using potassium salts of fatty acids
Moderately soluble in water	Easily soluble in water
Does not lather well	Lathers well

3 This substance killed thirty-six children in France in 1972, when a manufacturer of baby powder mistakenly sold a product containing 6 percent hexachlorophene, which was far too high a concentration. Following this incident, and the discovery that a further fifteen deaths in the United States were probably linked to hexachlorophene exposure, the substance was banned in several countries, including France and the US.

the fatty acids generally remains in the soap. This is not a bad thing, as it protects our skin from the high pH of the finished soap and can reduce dryness. Mass-produced formulations often intentionally use more fats than just those calculated to completely saponify with the lye so that the leftover oil in the finished soap makes it creamier and more hydrating. Another trick the industry uses to make soaps kinder to our bodies is replacing the water in the recipe (either in part or in full) with milk or even cream. The extra fat and protein act together to moisturize and soften our skin.

The hot process uses an excess of lye to force all the fats to react. The mixture is heated and cooked, sometimes for several days depending on the type of fat used. Then, a solution of sodium chloride is added, which causes the soap to separate from the excess lye. The precipitate (that is, the soap bars) is left to cure for several weeks before the bars can be used. This process is not usually done at home, primarily because soap with excess soda can be extremely harsh and irritating to our skin.

CLEAR SOAP

Some soaps are translucent, and others can even be transparent. It might seem like some far-fetched modern chemical wizardry, but the discovery that soap could be clear was actually first made in 1789. A chemist saw that when an opaque soap was heated with ethyl alcohol, poured into a mold, and left to cool, it turned transparent. Clear soap seems to have some sort of magical effect on consumers, perhaps because of the sense of purity and freshness it conveys.

Hard clear soaps are made with sodium hydroxide and liquid clear ones with potassium hydroxide. The distinctive feature of these recipes is that they contain ethyl alcohol as well as sugars or polyols such as sucrose, glycerol, sorbitol, and propylene glycol.

INGREDIENTS IN CLEAR SOAP[4]

INGREDIENT	QUANTITY
Castor oil	30 g
Coconut oil	45 g
Olive oil	15 g
Palm oil	60 g
Basic solution	
Deionized water	50 g
Sodium hydroxide	22.68 g
Sugar solution	
Deionized water	24.10 g
Sucrose	27.79 g
Other ingredients	
Ethyl alcohol	60.37 g
Glycerol	37.05 g
Perfume	3.73 g

4 Suzanne T. Mabrouk, "Making Usable, Quality Opaque or Transparent Soap," *Journal of Chemical Education* 82, no. 10 (October 2005): 1534–1537

Homemade soap

Over the past few decades, there has been renewed interest in the craft production of soap, usually using the cold method. Speaking as a chemist, I find making soap at home to be a hugely fascinating and instructive enterprise. If carried out with the necessary preparation and supervision, it can bring you closer to performing actual science than most other experiments. Indeed, it can even trigger a love of the subject in folks who hated science at school because it seemed too far removed from reality. There is no denying that this interest is shared by hundreds and thousands of people who post recipes, tips and tricks in books, online forums, Facebook groups, YouTube channels, and Instagram profiles all dedicated to the same passion: soapmaking.

However much I love the idea of doing some hands-on chemistry at home while observing one of the most ancient chemical reactions, I decided not to include an explanation of how to make soap in this book. I say this with a heavy heart, but home recipes include some very hazardous chemical substances (lye, for example), creating the potential for something to go wrong, especially for beginners who don't have much experience handling such dangerous substances and don't have a resident expert around. That said, even if you're more experienced, not every household has the necessary equipment to safely produce soap (like a digital thermometer and a pH indicator). Since I've written this book for the average reader rather than the qualified chemist, I fear that I might encourage someone to underestimate the risks and have a go at mixing dangerous substances.

One of the main issues with making soap at home is that it's difficult to know exactly how much lye you need to convert the fats to soap, especially if those fats are leftover cooking oil (which is quite common). Without careful chemical analysis to establish the exact composition of the oils, you risk putting in too much sodium hydroxide and producing a highly corrosive soap. Even if you use the same oil each time, the composition can vary a lot from year to year due to the sensitivity of vegetable products to climate and location. One practical solution often used in soap-making recipes, even industrial ones, is "superfatting": adding an excess of oil to the batch to make sure the lye gets completely saponified. Even with this tactic, the mixing process might not be perfect, so you can never be sure that there are no tiny pockets of unreacted lye. This is why homemade soap should always be rinsed before wrapping. All things considered, however, making soap in a school science lab under the supervision of a teacher or lab technician and following all necessary safety precautions is an excellent way of showing students how chemistry is all around us.

SOAP INGREDIENTS

I mentioned before that hard soaps are the sodium salts of the fatty acids in animal and vegetable fats, while liquid soaps are potassium salts. That said, soap's properties also depend on the particular fatty acids used and thus on the actual oils and fats they originate from. Until the twentieth century, both animal and vegetable fats were reacted directly with alkaline substances. Early soaps were usually produced from a single type of fat, like olive oil, mutton fat, or sheep fat. Animal fats were easily accessible and tended to be the first choice. As the production of vegetable oils like cottonseed, palm, and coconut expanded, soap recipes became more sophisticated, since a blend of balanced fats produces a much higher-quality soap with a wider range of properties. This is because fatty acid salts have varying characteristics: One lathers nicely while another doesn't, one is harsh on the skin while another is kinder, and one produces a crumblier soap while another holds together well. Palm oil, for example, creates a harder soap, but coconut oil makes more bubbles. Even the color depends on the type of fat: Butter results in a beige soap and sunflower oil a white one.

DID YOU KNOW?

The extent to which a soap lathers and its degree of stability do not necessarily correlate with how well it cleans. Nowadays, consumers tend to think that bubbles remove dirt, which is not exactly true—actually, they can sometimes be more of a hindrance. However, since people have learned to expect bubbles, manufacturers aim to reflect this requirement in their commercial soap formulations.

When soap is made industrially, whether for personal hygiene or laundry, a mixture of oils is almost always used. Traditionally, the soap industry splits oils into two types: nut oils (from hazelnuts, walnuts, coconuts, and palm kernels—the latter are the edible seed of the oil palm's fruit), which mainly contribute lauric acid to the soapmaking process, and non-nut oils (such as palm oil and animal tallow), which mostly provide palmitic, stearic, and oleic acids in different proportions.

The oils in each category lend different properties to soap. Tallow and palm oils produce soaps that are more or less insoluble at room temperature, making them practically latherless. They make the soap harder, which stabilizes any production of bubbles—any lathering that occurs normally comes from other oils in the mixture. Nut oils are more soluble in water and lather well, and they are also generally softer. Commercial recipes usually contain 70 to 80 percent tallow or palm oil and 20 to 30 percent coconut or palm kernel oil.

CHOOSING A SOAP

This book won't look at personal hygiene products for two reasons: Firstly, personal hygiene products are classified as cosmetic goods and subject to different regulations, and secondly, it would take another three hundred pages or more to fit it all in! Instead, we'll focus on the classic bar of laundry soap—although a lot of what you'll read here also applies to products for bathing. This should always be your rule of thumb: To know if a soap is right for you, don't just read about its composition, but also try it out so you can check its consistency, scent, and performance.

Does the type of oil matter?

Let's clear up some basic chemistry before we go any further. All soaps sold at the store are made from vegetable oils or fats. As far as I know, there are none made solely with animal fat. If there were, you'd find sodium tallowate, an animal fat, in the list of ingredients. In any case, after saponification, all of the oils (meaning the mixture of triglycerides) cease to exist—they disappear, converted into something else. These new molecules, the soaps, are no longer oils and no longer have their original properties. This may seem fairly obvious and pointless to repeat, but quite frequently, we subconsciously assume that the finished product has the same properties as the starting ones.

In theory, we could make the exact same soap with the same chemical composition by starting from different varieties of oil blends, precisely because the ingredients are always completely transformed. Dietary fats and oils are a mixture of triglycerides featuring the same sets of fatty acids in differing proportions. The three fatty acids that we humans most commonly accumulate in our adipose tissue are palmitic acid, oleic acid, and linoleic acid, the same ones found in dietary oils. Ultimately, a soap's properties depend on the mix of fatty acids, and it doesn't matter whether a specific fatty acid comes from olive oil, palm oil, or coconut oil.

This said, some soaps' labels boast that they contain wild and wonderful varieties of oils, from the classic olive to the less common almond to the more exotic argan or jojoba. A few very rare cases aside, please don't think that these oils are the soap's main component. You'll easily discover this by flipping the bar over and checking out the ingredients. At the top of the list will most likely not be the oils blazoned across the front, but rather palm or palm kernel oil, coconut oil or another similarly inexpensive oil, or saponified oils such as sodium palmate or stearate. Toward the bottom of the list, you'll find the oils named on the front label. This doesn't mean that these latter oils serve no purpose, though—as I explained

> **DID YOU KNOW?**
>
> If glycerol is not separated (usually by washing with water and salt), it remains in the final soap at about 6 percent of the product's total composition. Traditional Marseille soap is made by heating oils with lye for many hours, then washing with salt water to separate the glycerol from the insoluble soap.

earlier, sometimes excess oil is used to prevent any unreacted lye from being left over, and extra oil can add skin-moisturizing properties. However, don't forget that a soap's cleaning power depends on the oils converted during the soapmaking process, while any added extras are useful for creating a particular scent or simply setting a product apart for marketing purposes.

Are all soaps equal?

While it may be true that all oils contain the same fatty acids, we still need to remember that the proportions in which those acids are present vary. Olive oil has a very different composition of fatty acids than corn or coconut oil; therefore, soaps made from these different oils may have very different properties. The one thing we shouldn't do is assume that an oil's dietary properties (positive or negative) transfer directly when that oil is used to make soap. As I explained, the oil disappears during the saponification reaction. (As further proof: Oils can be eaten, but soaps can't!)

What is neutral soap?

Some soaps are said to be "neutral." In chemical terms, neutrality refers to a substance's acidity or alkalinity. Technically speaking, this would mean that a neutral soap has a pH of 7, the same as distilled water, which is impossible. All soaps are basic,[5] and if you don't believe me, measure the pH of water after you've dissolved some soap in it using a simple test strip or aquarium pH meter. On some soap packaging, there's an asterisk beside the word "neutral." I recommend that

PROBLEMS WITH SOAP

If you use soap regularly and leave it sitting on the side of the sink, you'll notice that it often goes soft. This happens if it's left sitting in contact with water because the moisture penetrates the soap and changes its chemical structure. Some soaps are more susceptible to this than others, depending on the composition of the oils and the amount of water left in the soap during manufacturing. Another problem that often occurs when soaps are left unused for extended periods is cracking: Moisture gets into parts of the soap and wraps around the solid crystals, causing the bar to expand. When this swollen, watery area dries, it then contracts, leading to the cracks we often see in the surface. Oilier soaps made from oleic acid are less prone to cracking, whereas a high glycerol content increases the chances of crevices forming.

5 In chemical terms, all salts obtained from weak acids (like fatty acids) through a reaction with a strong base (like lye) are basic.

you always check what it's there for! You might find that "neutral" simply refers to the fact there is no unsaponified lye. Thank goodness—a soap free of an aggressive caustic substance should be the least we can expect in terms of safety!

Adding to the confusion is the fact that some detergents sold as neutral, such as intimate washes for women, have very acidic pH values. How does this work? Well, in this particular context, the descriptor "neutral" has no chemical meaning. It refers to the fact that the product's acidity is similar to that of the skin or membranes it is intended to come into contact with. (Yes, our skin is naturally mildly acidic.) Overall, advertising language is more evocative than accurate or scientifically correct. No one would buy a product labeled "acid soap," would they? The mystery of some "neutral" soaps is often revealed when you read the ingredient list: Instead of being real soaps (that is, the sodium salts of fatty acids), they contain synthetic detergents, which we'll discuss in the next chapter.

Is it Marseille?

With the advent of synthetic detergents, traditional Marseille soap gradually disappeared from stores, although the name never quite died out. In recent decades, it has actually made a bit of a comeback in some countries—although in name only, as olive oil is almost always absent from the formulation. When this oil does appear, it's often in such a small percentage that it is irrelevant. Nevertheless, many Mediterranean countries remain quite attached to this classic soap, a fact that is greatly exploited by marketing departments.

DID YOU KNOW?

The Greco-Roman physician Galen of Pergamon (who lived in the second century CE) once described the medical applications of soap, particularly for skin afflictions. Until the end of the nineteenth century, it was a regular treatment for psoriasis, scabies, herpes, and other skin conditions.

Marseille soap has not yet been granted official PGI (protected geographical indication) status in the European Union, which would protect the name from anyone and everyone claiming it for their own product. Indeed, almost any soap on the market can be labelled Marseille, although I'll bet that many of these impostors won't have come anywhere near olive oil. The white version of "Marseille soap" is obtained mainly from palm and coconut oils—although, as we learned earlier, this does also happen with the blocks produced in Marseille itself. Only the green version is made with different variations of olive oil (while the original soap is required to contain exactly 72 percent olive oil). When you think about it, though, a bar of soap made almost entirely from olive oil would cost far more than most of us would be willing to pay.

Basically, if a bar of soap has the word "Marseille" printed on it somewhere, it just means that it's a fairly hard laundry soap. It might contain, for example, sodium palmate, water, sodium stearate, glycerol, sodium palmitate, sodium chloride, perfume, sodium etidronate, or limonene. (At this point in the book, I'm fairly confident that chemical names won't make you scared of your soap, since you're learning that each substance is typically added for a reason.)

The legend of Mount Sapo

It's not completely clear where the word "soap" came from, and as you'd expect, there are many legends about its origins. One in particular alleges that soap was named after a mountain called Mount Sapo near Rome, where ancient Romans used to sacrifice animals as burnt offerings. Fat from the animals would mingle with ash from the fires and harden on the clay which, as the story goes, found its way into the Tiber River when it rained. This primitive soap dissolved in the water, and the local washerwoman who went down to the river found that the cream-colored water with hard blocks floating in it got their laundry cleaner—and thus soap was born. However, this story is probably completely fictional, as there is no such thing as a Mount Sapo near Rome.

5

Detergents

Soaps are great, but they come with some downsides. The first is that they don't clean very well in cold water because they're not soluble at low temperatures, but they must dissolve to have any effect.

Another flaw is that soaps don't work as effectively in water with a high mineral content (so-called "hard water"), especially in the presence of calcium and magnesium chlorides. This became more noticeable during World War II, when soldiers were forced to bathe in seawater. Since the calcium and magnesium salts in soaps are insoluble, soap used in salt water leaves behind a precipitate on skin and clothing. After multiple washes, these precipitates turn fabric gray (in the same way they deposit on surfaces—leaving a difficult-to-budge ring of dirt in the bathtub, for example). Besides, more soap than usual is required for washing in hard water because the calcium and magnesium present in the water react with it, using it up.

The third downside to soaps is that they are relatively basic, with a pH between 9 and 11. This is great for removing problematic oily stains in the short term, but over time, it can damage more delicate fabrics. Worse still, high alkalinity can irritate your skin. Skin naturally has the pH of a weak acid (somewhere between 5 and 6.5), and soap strips away its protective acid and sebum mantle. This is only temporary, however: In most cases, skin bounces back to its normal acidity fairly quickly. But sometimes, soap can cause dryness and irritation, or occasionally quite serious skin problems—not to mention that it can interact badly with the delicate, moist inner linings of some organs and body cavities.

Similarly, it's not a good idea to wash your hair with soap, as its basic pH causes the cuticles to rise along your hair shafts, leaving your hairdo brittle and frizzy. This is why our grandparents and great-grandparents—in the absence of modern shampoos, which have neutral or mildly acidic pH values—used to douse their hair in water and vinegar after using soap. Acetic acid neutralized the remaining alkalinity and flattened the cuticles to leave their hair looking glossy and smooth again. Nice hair, but with the distinct odor of a bag of salt and vinegar chips! Although I personally appreciate not having to wash my hair with soap and enjoy the benefits of shampoo that contains detergent, I still keep a 2- to 5-percent citric acid spray in the shower just to guarantee that perfect smooth and shiny finish. (Yes, even when I'm washing my hair, I'm still a chemist.)

Going back to the past, the fats and oils used to make soap fell into short supply during World War II. At the same time, automatic washing machines became more popular just as powder detergents were coming into vogue—all of which explains why laundry soaps very quickly vanished almost entirely in the 1950s and were replaced by detergents.

WHAT ARE DETERGENTS?

In everyday parlance, we normally use the word "detergent" to refer to a product that cleans. This would seem to make soap a detergent. But in chemical terms, as we've learned, soap is actually described as the sodium and/or potassium salt of one or more fatty acids, and it's classified as part of a family of molecules called surfactants. Scientifically, a detergent is a synthetic surfactant that is not a soap but is good for cleaning. Laundry cleaning products (colloquially called "laundry detergent") contain a detergent in combination with an assortment of other molecules required to balance the formulation.

In this book, depending on the context, the term "detergent" will refer both to its technical definition as a synthetic surfactant and to its role as a cleaning product containing synthetic surfactants. If you ever happen to pick up a product that claims to be a "detergent soap," you'd better look at the list of ingredients to understand what's really in it.

Surfactants

A surfactant is a molecule with two parts, each behaving differently: One loves water and seeks it out (so it's called "hydrophilic") while the other hates and avoids it (so it's known as "hydrophobic"). One way the hydrophobic part evades water is by inserting itself into a fat, since fat is also hydrophobic—birds of a feather flock together!

In low concentrations, all surfactants exist as separate molecules spread out in water. The higher the concentration, the more the molecules assemble into loosely bound spherical shapes called micelles, which protect the parts that seek to avoid water. In each micelle, the hydrophobic tails point toward the core, while the hydrophilic heads point outward. At even higher concentrations, micelles form more complicated tubes and layered structures.

To help visualize what a surfactant is, I have been using soap as an example. Not all surfactants are structured with heads and tails, but all of them (even the synthetic ones

PARTS OF A SOAP MOLECULE

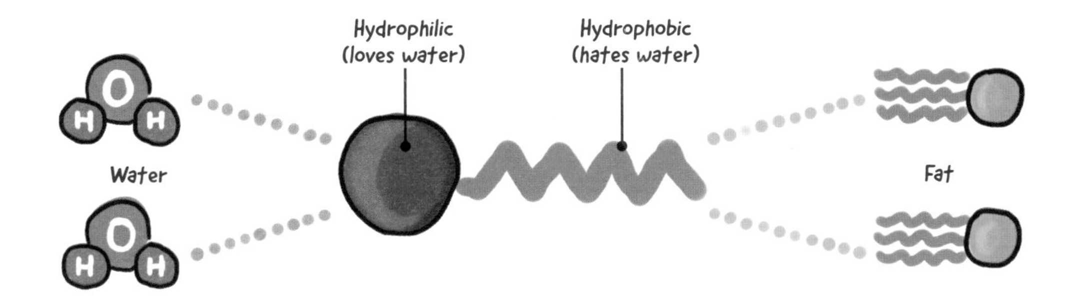

that have almost completely replaced soaps) have a part that prefers water and another that prefers to be in contact with fat (called "lipophilic").

How does a surfactant clean?

Now, let's explore how surfactants work and why they can be exceptionally useful when it comes to cleaning. A surfactant's molecules have multiple properties, and they simultaneously perform a number of equally important functions. We will use the process of cleaning a piece of fabric as an example.

WETTING THE SURFACE

This might seem obvious, but if a detergent is going to help water to clean a piece of fabric, then the water needs to get onto the fabric's surface and penetrate its fibers before it can even begin to remove the dirt.

To visualize this, put a drop of water in a plastic bowl. If you look closely, maybe through a magnifying glass, you'll see the bead of water trying its best to avoid contact with the bowl's surface. Water, by definition, does not like hydrophobic substances and seeks to touch them as little as possible. If there were no gravity, the bead would probably lift up into a perfect sphere. Here on gravity-bound Earth, the bead has to compromise and spread out somewhat, just enough to form the smallest possible surface area. I won't go into too much detail about the underlying physics except to say that this occurs due to a phenomenon experienced by all liquids called "surface tension."

Surface tension explains the fairly dramatic tendency of a liquid to resist changing its surface area.[1] Water has a fairly high surface tension, but it can be lowered significantly by adding a surfactant, which enables water to spread out more easily across a surface by reducing the angle of contact. In our fabric scenario, this means more of the water's surface area comes into contact with the textile fibers. The fabric then gets wetter (referred to as "increased wettability").

If you replace the water with oil in the plastic bowl experiment, you'll see that the drop spreads out much faster. Oil has a lower contact angle because its molecules behave in a similar way to plastic molecules. This also explains why it's harder to clean grease from a plastic surface than from glass or metal.

LEVELS OF SURFACE TENSION

(α = angle of contact between water and surface)

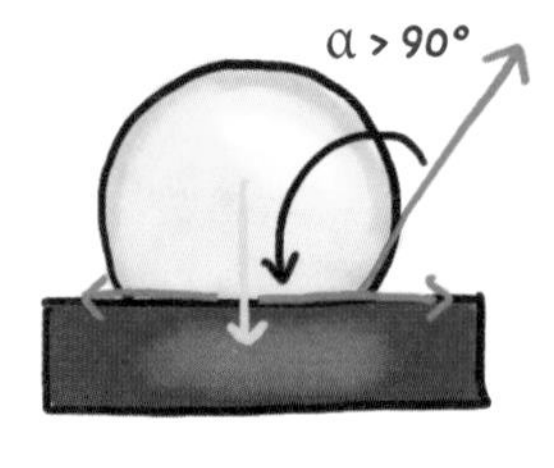

Hydrophobic surface

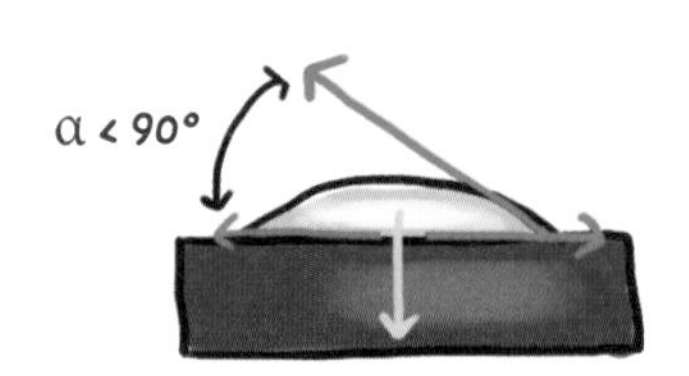

Hydrophillic surface

α = 0°

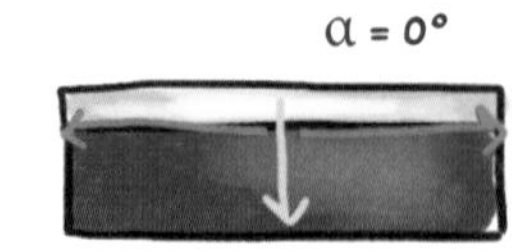
Perfect wetting

1 For a given volume, a sphere is the shape with the smallest surface area.

SURROUNDING THE DIRT

If grease has penetrated the textile fibers of our imaginary fabric, the hydrophobic tails of the surfactant reach into the stain and begin to lift up the grease. In technical speak, we call this the "roll-up effect," and it is key to surfactants' ability to remove dirt. Bit by bit, as more and more surfactant molecules manage to squeeze in between the fabric and the grease, the dirt gradually detaches from the surface, helped along by the mechanical effect of scrubbing. Dirt is much easier to detach and wash away when its contact angle with the surface is higher than 90 degrees.

FORMING MICELLES AND EMULSIFYING THE DIRT

If the concentration of surfactant is high enough, with the hydrophobic tails striving to stay away from the water and the hydrophilic heads trying to interact with it, spontaneous structures called micelles assemble. The tails cluster together into a spherical structure, with only the hydrophilic heads touching the water.

These structures play an important (albeit secondary) role in removing dirt. Once the grease is pulled away from the fibers, it ends up at the center of the sphere. It's held in suspension, meaning it exists in solid particles separate from the water, and emulsified, meaning the grease and water are combined despite their tendency not to mix. This is similar to how the molecules of lecithin (a surfactant) in egg yolks keep the oil droplets suspended in a solution of water and lemon juice (or vinegar if you prefer), otherwise known as mayonnaise. Micelles also help to break the particles of dirt into smaller fragments.

SURFACTANTS REMOVE DIRT

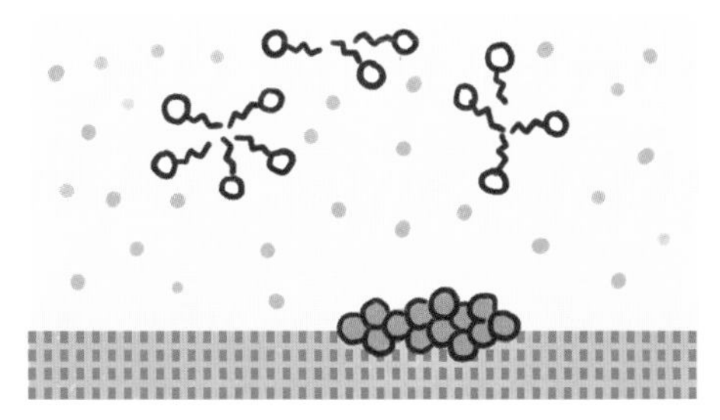

Surfactants form structures called micelles that separate the dirt from the overall solution

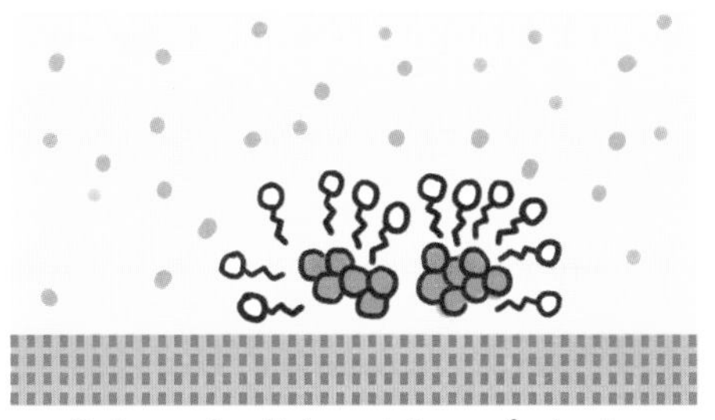

Detergent, which contains surfactants, wets the dirty surface

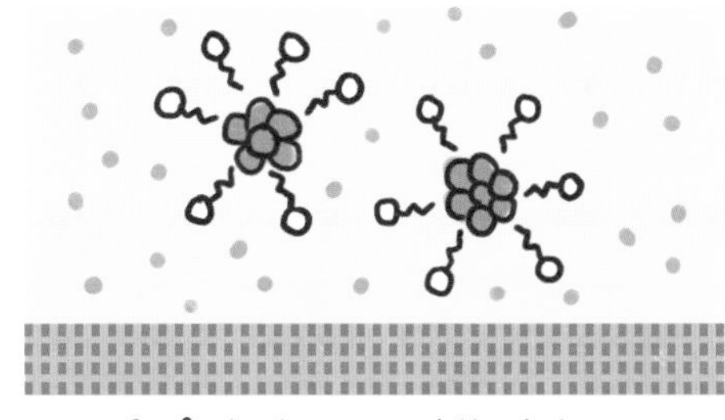

Surfactants surround the dirt

Dishwashing soaps clean dirty dishes using this same ability of surfactants to form emulsions and hold droplets of oily dirt in suspension. The high concentration of surfactants in these detergents means they act as an interface between the dirt and water, lowering the surface tension and creating an emulsion that carries away the dirt.

THE HISTORY OF DETERGENTS

In the beginning there was only soap. Historians and archaeologists have found multiple traces of its use by ancient civilizations: the Babylonians in 2800 BCE, the Sumerians in 2200 BCE, the Egyptians in 1500 BCE, and the Phoenicians in 600 BCE. As we saw earlier, theirs was a primitive soap made by boiling the ash of an alkaline substance like potash or soda with animal or vegetable fat. This method continued with only minor adjustments until the eighteenth century, and the resulting soap was used for bathing, laundry, and even medicinal purposes.

As the nineteenth century unfolded, powder or flake soaps began to appear. Chemists and industrialists experimented with new products for household cleaning, laundry, and personal care. In 1878, the German company Henkel launched a product called Bleich-Soda, a mixture of sodium carbonate and sodium silicate meant to be used alongside soap. It was an early limescale inhibitor designed to prevent the buildup of annoying insoluble calcium deposits that soap left behind as a precipitate when used in water.

In 1907, the first-ever detergent containing bleach was introduced, again by Henkel. It was called Persil, and it combined sodium carbonate, sodium silicate, and sodium perborate (hence its name: per (perborate) + sil (silicate) = Persil). Thanks to this product, sheets no longer had to be bleached dry in the sun for days on end.

The first synthetic detergent was created by chance in 1831 when French chemist Edmond Frémy heated sulfuric acid and olive oil. When he poured the dark brown mixture into water and neutralized it with lye, he discovered that the resulting concoction behaved like soap: It lathered and removed oil from greasy things. However, it took another hundred years after Frémy's discovery for a detergent containing synthetic surfactants to appear on the market. Further progress was made in the manufacture of soap-like molecules in 1916, when chemist Fritz Gunther obtained a product called "Turkey red oil" from castor oil and sulfuric acid.

The first synthetic laundry detergent was probably Dreft, introduced by Procter & Gamble in 1933, followed by Fewa from Henkel in 1935.[2] One of Procter & Gamble's researchers, Robert Duncan, had visited the research labs of two German chemical conglomerates. At IG Farben, Duncan was shown a "wetting agent": a substance that enables water to wet textile fibers more effectively so they can be more evenly dyed. As we learned earlier, surfactants like soap have this wetting effect. The chemists at IG Farben had learned of a textile company that used a bile

2 Some sources cite 1932 as the year Fewa was first marketed, which would mean Fewa actually holds the record for the first synthetic detergent ever sold.

extract instead of soap to perform this function and had succeeded in reproducing the same substance in their laboratory. They marketed it under the name Igepon. At the second chemical company Duncan visited, scientists from Deutsche Hydrierwerke (later acquired by Henkel) showed him a product they had synthesized to compete with Igepon. Duncan had one hundred kilograms (about 220 pounds) of it shipped to his head office in Cincinnati for testing, and the following year, Procter & Gamble launched Dreft, the first detergent sold in the United States.

Unlike ordinary soap, this early powder detergent didn't leave any nasty insoluble calcium deposits on clothes. That said, it struggled with more stubborn dirt. The real turning point came when Procter & Gamble chemists combined their synthetic surfactant with a substance that controls calcium: sodium triphosphate. The resulting product, named Tide, launched in 1946 and was an immediate hit: It took over 30 percent of the American market in only a couple of years, and an updated formulation continues to be the best-selling detergent in the US.

The soap that wasn't a soap

Synthetic detergents initially found success as laundry cleaning products, but they soon began to replace personal soaps as well. They continue to dominate this market area in shampoo, bubble baths, and the liquid soaps we inaccurately refer to as "hand soap."

Persil, Henkel, 1907

Dreft, Procter & Gamble, 1933

Tide, Procter & Gamble, 1946

Synthetic detergents' rise to prominence began in 1955, a watershed moment in the chemistry of detergents—the year Dove soap was unveiled. For the first time, Unilever had succeeded in adding synthetic detergents to their classic soap recipe, producing something that wasn't really soap anymore. "Suddenly Dove makes soap old-fashioned!" read the slogan. This new beauty bar had a much gentler pH than normal soap and didn't leave nasty rings in the bathtub.

> **DID YOU KNOW?**
>
> Anionic surfactants introduced after World War II were primarily obtained from petroleum-based sources. More modern nonionic surfactants are mainly manufactured from vegetable sources like palm and coconut oil.

Marketers' habit of calling something "soap" when it's really not is still prevalent today. The next time you head to the grocery store, have a look at the liquid soaps and check the ingredients on the back—I guarantee they'll be synthetic detergents. There is such a thing as real liquid soap, but it's not used for cleaning anymore.[3]

Beyond laundry and personal hygiene, the third area in which synthetic detergents have found success is the home, a place with many different surfaces to clean and many different types of dirt. From bathrooms to kitchens, getting a window streak-free and sparkling clean is an entirely different job from cleaning a stovetop, a parquet floor, or a toilet bowl. Therefore, it's not only sensible but also desirable that we have access to a variety of different products designed for different combinations of surfaces and dirt. Then again, you could also say that kitchen and bathroom floors are about the same, as are windowpanes, glass shower panels, and mirrored wardrobe doors—which is why there's such a thing as a multipurpose cleaner.

SOAP AND DETERGENT pH VALUES

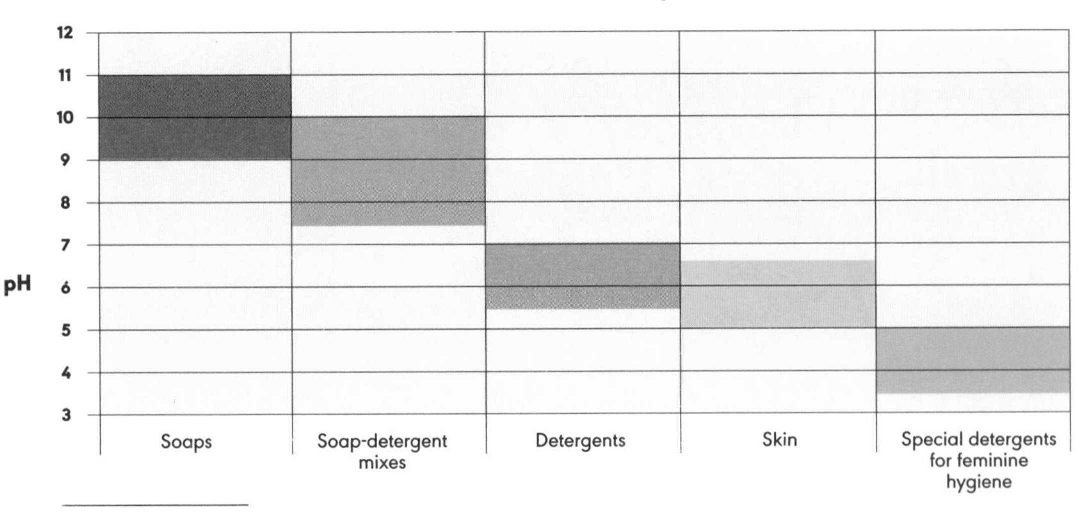

3 Some liquid soaps are still used as agricultural pesticides that are sprayed on plants.

Surfactant classification

Surfactants are generally classified according to the charge of their polar head group (the part that loves water). Anionic surfactants have a negative charge, cationic surfactants have a positive charge, nonionic surfactants have no charge (although they do have a small area that's drawn to water), and amphoteric surfactants have both positive and negative charges.

Each type has specific properties that make them suited to particular uses. Anionic surfactants, for example, are the primary go-to for laundry products, with nonionic surfactants in second place. Amphoteric surfactants are not widely used, whereas cationic surfactants appear in a number of special products. Dishwasher tablets and capsules only ever include nonionic surfactants on account of their excellent cleaning power combined with minimal foaming to avoid damaging the appliance.

ANIONIC Anionic surfactants are the oldest, with the forefather of the category being soap itself. They are the preferred choice for both hand- and machine-wash laundry detergents, and they are most likely to be made from sulfates[4] or sulfonates.[5] This type of surfactant is generally inexpensive, an effective detergent, a good emulsifier, and very foamy.

The most popular surfactants used in laundry products are alkylbenzene sulfonates.[6] Unlike ordinary soaps, their ability to tackle grease is less affected by water hardness since their calcium and magnesium salts are soluble. They also work well in cold water, but they can irritate the skin more than other surfactants, so when they're used in hand-washing products, they need to be combined with additional substances. One anionic surfactant that's commonly used in shampoo as well as hand and body wash is sodium lauryl sulfate.

Detergents often feature a combination of surfactants to prevent worse performance in hard water and, in the case of hand-washing, to ensure a decent foam.

NONIONIC Nonionic surfactants make great detergents, even in hard water. They are also much softer on delicate fabrics and less harsh on the pigments and dyes used to color textiles. Ethoxylates are the most common type in this category. Nonionic surfactants tend to be better at removing greasy dirt than anionic surfactants and are less irritating to skin.

CATIONIC Cationic surfactants are much less common—their main applications are in fabric softeners or delicate detergents designed for wool. They're less useful for cotton because cotton fibers usually take on a negative charge in water, meaning they are naturally attracted to the positive charge on the hydrophilic ends of cationic surfactants. Also, at low concentrations, these surfactants reduce the repulsion between dirt and fabric fibers, making them less suited for cleaning. The most common cationic surfactants are quaternary ammonium compounds.

AMPHOTERIC This rarely used last group contains surfactants that are both anionic and cationic. They are often considered a good choice since they're effective cleaners that are delicate on the skin, but they are more expensive than other options. Betaines are frequently used amphoteric surfactants.

---

4 The general chemical formula for sulfates is $R\text{-}O\text{-}SO^{-3}$.

5 The general chemical formula for sulfonates is $R\text{-}SO^{-3}$.

6 In particular, those known as linear alkylbenzene sulfonates (LAS).

WHAT'S IN LAUNDRY DETERGENT?

The first successful laundry powder, Tide, contained nothing more than a synthetic detergent and a substance to prevent limescale deposits—but its composition changed notably in the following years. Both machine- and hand-wash laundry detergents added multiple new ingredients, each serving a specific purpose, to keep up with consumer expectations and changes in the way people did laundry.

Sticking to the basics, let's have a look at what's in laundry detergent. This will help you to better evaluate the labels at the grocery store and understand that not all detergents are equal (unlike what some people might claim). With these two advantages, hopefully you'll be better at picking the right product for you.

Some ingredients are found only in powder detergents, others only in liquid detergents, and some in selected products but not in others—sometimes for financial reasons, sometimes for technological ones.

Surfactants

The key ingredient in laundry detergent is the surfactant. Since surfactants are relatively expensive, the proportion varies from product to product—from more than 50 percent in premium products to less than 5 percent when the surfactant serves solely as a wetting agent and the task of cleaning is left to the water.

Nowadays, a mixture of surfactants is often used, with anionic making up the lion's share of both powder and liquid products. Anionic surfactants perform better than soap in hard water but are still susceptible to its effects, so they are less useful in places where the local water has a high mineral content. In hand-wash detergents, they can also irritate your skin. For these reasons, they are often combined with nonionic surfactants, which perform well in soft or hard water and are kinder to skin. They are also more compatible with other key ingredients—like enzymes, for example. The main disadvantage of anionic surfactants is their high cost.

Water softener

After the surfactant, the next most important ingredient is water softener. Water with a high calcium and magnesium mineral content impairs a detergent's cleaning ability because some of the surfactant (especially for anionic surfactants) binds with the ions in the water. Also, the resulting limescale buildup can damage your washing machine. (Earlier in the chapter, we saw how this was the main flaw that caused soap to be replaced.)

The first water softener was sodium carbonate (soda ash): Added to water, it precipitates the calcium salts as insoluble calcium carbonate. Sodium carbonate also lends alkalinity to water, which is essential for removing dirt. Even nowadays, this ingredient is still used on its own—although mostly in cheaper detergents—but the risk is that the calcium carbonate it creates can deposit on the clothes you're trying to clean, creating yet another problem. It's more commonly used in combination with other water softeners and as a source of alkalinity.

Sodium phosphate is highly effective as a water softener, and it was once widely used.[7] However, it's also very bad for the environment, causing excessive algae growth when discharged into rivers and streams, so it's been banned or tightly regulated in a number of places around the world, including the United States and the European Union.

In many countries, phosphates were replaced by zeolites, specifically zeolite A. Zeolites are inorganic, synthetic substances composed of silicon, aluminum, and oxygen. They are porous, which helps to trap calcium ions and release sodium ions in their place. However, zeolites are less effective at dealing with magnesium ions, and unlike phosphates, they have the added disadvantage of being insoluble. Therefore, the danger is that they'll deposit on the clothes you're washing or, worse still, that carbonate crystals will form in the fibers, leaving the fabric stiff as a result. For these reasons, zeolites are always used in combination with other ingredients.

Citric acid, which converts to citrate in the alkaline environment of a detergent, can serve as a water softener because it binds to minerals to prevent limescale from forming. It is found in many liquid detergents. Polycarboxylates are water-soluble polymers that perform this same function. Other products may contain soap for precipitating limescale.

STRUCTURE OF ZEOLITE A
Imagine hundreds of these arranged one beside the other

Polymers

If you look at a detergent label, you'll most likely spot an ingredient with the prefix "poly-" (polyacrylate or polyacrylic acid, for example). These are polymers: very long molecules made of multiple smaller chemical units linked together in chain-like formations. In a detergent, these molecules disperse dissolved dirt to prevent it from redepositing on textiles. A wide range of polymers are used for this purpose, including those made from cellulose (for example, carboxymethyl cellulose, or CMC), which is a natural polymer made up of glucose molecules that's also the main substance in plant cells. Some polymers are added to detergents to prevent dye transfer and color staining of fabrics in the wash. The specific polymer required depends on the type of dirt that's being removed, which is a choice the manufacturer makes when defining their formulation.

7 In particular, sodium tripolyphosphate (STPP) is so effective that it has remained the benchmark for water softeners.

Enzymes

Surfactants improve water's ability to clean, but only to a certain point: They work better on some molecules (oily stains, for example), but are less effective on others (like proteins and starches). For this reason, almost all modern detergents (not just laundry detergents) contain a particular type of molecule that is becoming increasingly important to detergent formulators: the enzyme.

Enzymes are special molecules that, depending on the type used, can target and break down protein, starch, or lipid deposits into smaller fragments that can then be carried away by surfactants. Being proteins themselves, enzymes also biodegrade very easily—a key trait at a time when the environmental impact of the products we consume is a growing concern. It's predicted that enzymes will play an increasingly important role in the years to come, and new ones will be developed alongside those already in use.

PROTEASES

German chemist Otto Röhm was the first to explore (and patent in 1913) the use of enzymes for cleaning. Back then, proteases (which break down protein) were the only ones available, and they had to be extracted from the pancreases of dead animals. They were soon found to be too sensitive to the alkaline pH of cleaning products and were abandoned. The turning point came in 1959, when proteases were successfully produced from bacteria[8] that were resistant to alkaline pHs and could withstand temperatures of up to 150°F (or 65°C), a high regularly reached by washing machines across Europe at the time. That same year, Bio 40, the first detergent containing enzymes, was released in Switzerland.

Proteases are still the most common class of enzymes found in detergents.[9] These molecules—which are also naturally present in our intestines, performing the same function—break the peptide bonds in protein stains, splitting them into smaller, more soluble molecules that can easily be carried away by detergent. Protein is a component of many different types of dirt—including blood, food deposits (like eggs and milk), sweat, and grass—which is why this type of enzyme has risen to such prominence.

AMYLASES

Hot on the heels of proteases came amylases, which were used starting in 1973.[10] This class of enzyme attacks the starch in food residues, and I'm not just talking about pasta and rice: cocoa, for instance, contains a high percentage of starch, and removing it effectively means attacking the colored part of the dirty brown stain on your shirt. Amylases can do this because the starch is where the colorant has penetrated.

LIPASES

Lipases are another potentially useful class of enzymes, first added to detergents around 1988, when a cost-effective method of

8 *Bacillus subtilis* initially, then *Bacillus licheniformis*.

9 Whenever you see a molecule name ending in "-ase," it means that it's an enzyme: a protein that causes a specific chemical reaction to take place.

10 More specifically, α-amylases.

HOW THE ENZYMES IN A DETERGENT WORK

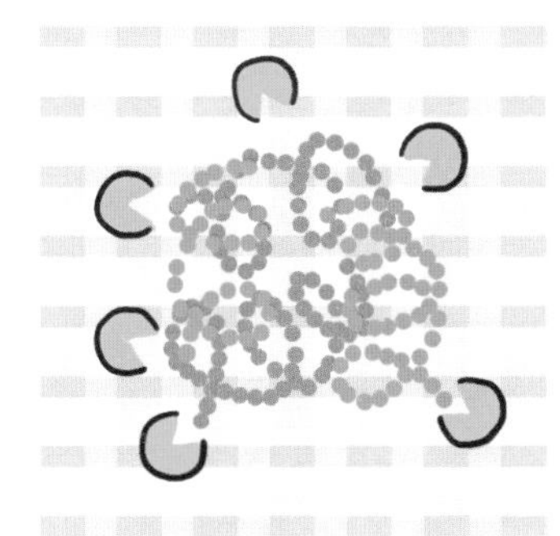

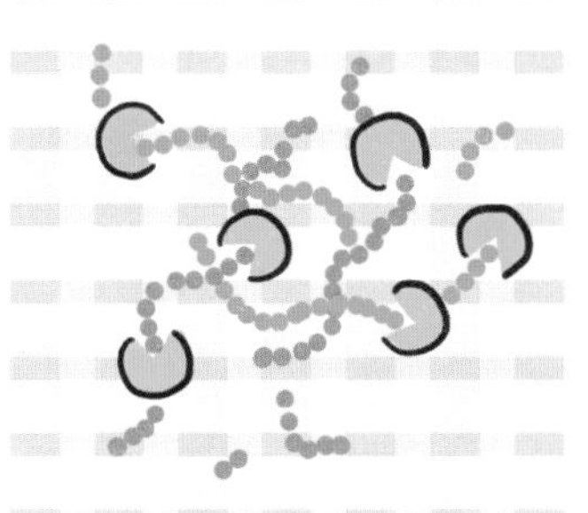

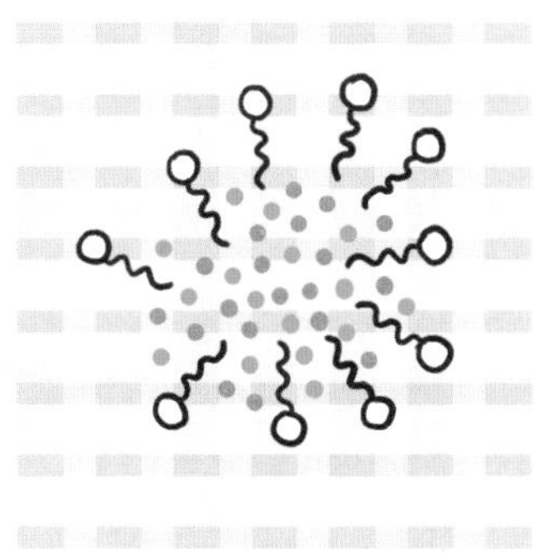

Enzymes target deposits of proteins, starches, or lipids

Enzymes break these deposits into smaller pieces that are easier to remove

Surfactants carry the smaller pieces away

producing them from genetically modified organisms was developed. The prefix "lip" denotes the molecule they target: lipids, or fats. Lipases split triglycerides into their component fatty acids, and they are especially useful for detergents that would otherwise struggle to remove oily dirt in cold water.

CELLULASES

The last group of enzymes to come onto the detergent scene was cellulases. Unlike their predecessors, these molecules do not attack dirt directly. Instead, they act indirectly on the cellulose in the fabric, removing a thin layer from the surface at the end of the cotton fibers, an area where dirt sticks.[11] The main function of cellulases is removing the fuzz and pills that form on cotton clothes.

OTHER ENZYMES

More enzymes are being developed or have been recently introduced: mannanases, for example, which break the structure of polysaccharides (galactomannans and glucomannans) found in the kinds of vegetable gums (like guar or locust bean) that are used to make ice cream. Theoretically, they will be able to remove ice cream stains that classic detergents have failed to budge. Many of these enzymes are now obtained from genetically modified bacteria or yeast.

Bleaching agents

Bleaching agents are chlorine and its related compounds (such as sodium hypochlorite) and peroxygens (such as hydrogen peroxide and sodium perborate). They are often an ingredient in detergents, and their purpose is destroying or modifying the color molecules in stains like those made by coffee, wine, or tomato sauce. I'll go into a lot more detail about the most common ones (chemicals like bleach, hydrogen peroxide, and hydrogen peroxide–releasing compounds, for instance) in the next chapter. All we need to know for now is

11 For any chemists reading: They hydrolyze the β-1,4 glycosidic bond.

that hydrogen peroxide–releasing substances have been an ingredient in powder detergents for almost a century. The first bleaching agent ever used, sodium perborate, made its debut nearly one hundred years ago.

Chlorine-based bleaching agents can damage colored fabrics, so they're not used in laundry detergents. Oxygen-based products like sodium percarbonate—which replaced sodium perborate because it was cheaper, easier to produce, and not as bad for the environment—are used instead. However, for percarbonate to be effective at low temperatures, an additional molecule is required. In the United States, this is normally the bleaching activator sodium nonanoyloxybenzenesulfonate (NOBS).

Bleaching agents are usually added only to powder detergents, as they are not compatible with liquid formulations.

Sequestering or chelating agents

Some metal ions are the nemesis of detergents. These ions can ruin their cleaning power, degrade their molecules, hinder their brightening potential, and even form insoluble precipitates on fabric surfaces, discoloring them. Iron can cause yellowish staining, for example. These metals can be found in water and dirt, and a huge array of naturally occurring color molecules contain substances like iron and copper.

To prevent any of this from happening, any metals that are present need to be captured. This is currently done using chemical substances that chemists call "sequestering" or "chelating" agents. "Chele" comes from Greek and means "claw," suggesting that the chelating molecule grabs the metal ions and captures them like a crab pinching them in its claw.

Two of the most common sequestering agents added to detergents and cosmetic products are abbreviated to EDTA[12] and DTPA.[13] They have been banned in some countries due to their poor biodegradability and harmful impact on the environment. Another example is HEDP, a chelating agent also referred to as etidronic acid, which is used in detergent formulations and occasionally for treating specific medical conditions. Citric acid salts can also serve as sequestering agents.

Optical brighteners

Depending on what a detergent is expected to do, manufacturers can choose to add a variety of additional components. Usually these appear in very low percentages, but they are important nonetheless. The most common among these are probably optical brighteners.

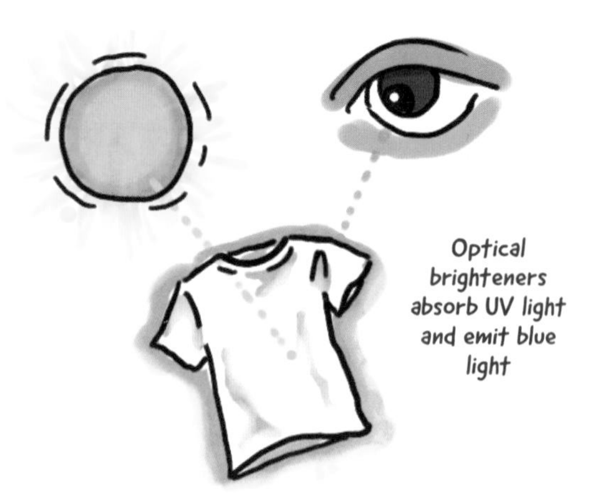

12 Ethylenediaminetetraacetic acid.

13 Diethylenetriaminepentaacetic acid.

These chemical compounds, which have been in use since the 1940s, absorb light in the ultraviolet and violet regions of the electromagnetic spectrum and re-emit light in the blue region.[14] The effect this creates can be seen with laundry that tends to appear yellowish after regular washing. Once an optical brightener is deposited on fabric fibers, the blue light emitted by these molecules makes clothes and sheets appear whiter and brighter by offsetting the yellow hue. This is where the name "brightener" comes from: They do not remove the substance causing the yellowing as a bleaching agent would; instead, they emit a light that suppresses it. Of course, this only works if the light shining on the clothes contains UV rays (like sunlight). Incidentally, optical brighteners are what give powder detergent the whiter-than-white appearance that was much touted in so many advertisements in the past.

You might be thinking, "What kind of hocus-pocus is this?" and in a way, you'd be right. Bear in mind, however, that untreated cotton is nothing like the dazzling white it becomes by the time we buy it as a nice new T-shirt. Natural cotton is either dyed to make it ultra-white or has optical brighteners added from the get-go. Sadly, these compounds don't last forever—they gradually wash away, which is why they need to be restored (at least partly) by a detergent. When optical brighteners are used to wash lighter or pastel-tone clothes (basically, anything that's not strictly white), the blue-violet light can actually make it look like the fabric is a completely different color.

Brighteners are also used in the paper industry to make paper look whiter and brighter by emitting more visible light than it absorbs. There are literally thousands of different compounds available that can absorb UV light and re-emit it as visible light of varying frequencies. Cosmetics is another area in which optical brighteners are used for purposes like camouflaging wrinkles and correcting hair yellowing.

Although adding optical brighteners to a fabric detergent might seem like scientific subterfuge—even deceit on the part of the modern chemical industry—in truth, it's actually just the evolution of a product that was in use from the mid-1800s to the mid-1900s. Back then, a small cloth pouch of solid blue dye would be added to a laundry load during rinsing.[15] Blue and yellow are complementary colors, which means that adding blue dye to a yellowing fabric makes it look much whiter. People knew that a trace of the dissolved blue dye would be enough to neutralize the yellowing of the fabric. These little pouches or tablets were called laundry bluing or dolly blue, and they came from a variety of brand names. Many are still on sale today if you want to try one.

Fragrances

Some people like to pop a few drops of essential oil into the washing machine. Manufacturers have tried the same thing and were faced with the same problem: Essential oils (those you can buy at the store, that is) do not dissolve in water.

14 Optical brighteners are often derived from stilbene (for example, 4,4′-diamino-2,2′-stilbenedisulfonic acid).

15 Often synthetic ultramarine blue or a variant.

To demonstrate this, try dropping oil into a glass of water—the oily drops just float around on the surface. In the washing machine, the drum's mechanical action may break them down into smaller and smaller droplets (depending on the type of cycle you pick), but since essential oils are not designed for this purpose, they won't emulsify in the wash water and won't spread through the clothes evenly the way commercial formulations can. As a result, most of them will end up being drained away. Essential oils are also allergenic, so they need to be handled with care. To find out more about detergent fragrances that *do* stick around on clothes, see the box below.

Other ingredients

Biocidal ingredients are often added to extend a detergent's shelf life. They're not always necessary in powder detergents, but they're crucial in liquid ones for preventing the proliferation of bacteria and mold, especially after the bottle has been opened. They're used across the spectrum of liquid cleaning products, from hand soap to shampoo.

Any other ingredients depend on the specific properties the manufacturer has in mind for their product. Foam control agents are one example: They're added either to increase the formation of foam or to hinder it so it does not damage an appliance. Soap is generally used for this purpose.

Thickeners are added to slow a detergent's flow, which can help with carefully measuring a pretreatment for laundry before it goes into the machine. Finally, liquid products typically contain colorants to create the same whitening illusion as the laundry bluing we discussed before.

AT THE GROCERY STORE

Laundry product scents

Violet, lavender, musk, and pine are just some of the many fragrances people like for their laundry. You can even buy products with a "fresh laundry" smell: that crisp, clean scent of laundry drying in the sun. How many detergent advertisements have you seen over the years that have waxed lyrical about this smell as their unique selling point? Most of us now instinctively associate clean clothes with a pleasant scent, and the cleaning industry adds fragrances everywhere, not just to laundry detergents. The compounds they use are complex and can feature a staggering array of different molecules, all blended to create a given fragrance. Producing a fragrance for a detergent isn't an easy process: The resulting blend has to dissolve in water (in both the bottle and the washing machine), not react with the other components in the detergent, have a specific pH, and be safe, biodegradable, stable, and not allergenic. That said, most of the perfume molecules end up going down the drain. Very few actually deposit on the fabric, and what little of the fragrance is left behind disappears pretty quickly. I am not a fan of these fragrances, I'm afraid, so I can't say any of this bothers me. In fact, I buy fragrance-free whenever I can. Nevertheless, the added perfumes can often be useful for covering up the otherwise unpleasant smells of the rest of the ingredients.

CHOOSING A DETERGENT

Back in the day, people used a wooden washboard and a bar of soap to do their laundry. Later on, a lucky few might have had a wooden tub of hot water in their home, but the majority cleaned their clothes in cold water at the communal washhouse or even in the river as many women in developing countries still do.

The main difference in cleaning materials used to be whether you could afford soap or had to use ash that you'd collected during the winter. Nowadays, we're treated to such a bewildering array of products at the grocery store that it's almost impossible to choose. Soap bars are still around, even if they're relegated to a dark corner, but we still have to navigate our way among powders, tablets, liquids, and capsules. Then there's our choice of detergent type: for delicates, colors, or whites?

Let's take a closer look at each of these decisions.

Laundry soap bars

As I mentioned before, you can still buy laundry soap bars nowadays. On closer inspection, the list of ingredients reveals that there are basically two types of bar: real soap (which almost always is marketed with connections to the city of Marseille, even if it was produced thousands of miles away and has nothing in common with the original Marseille soap) or soap mixed with synthetic detergents to ensure better performance in hard water. These days, soap bars are only really used for dabbing on the odd stubborn stain.

Powder or liquid detergent?

Walking down the household cleaning products aisle of a modern grocery store, you'll no doubt see row upon row of big, bold, brightly colored bottles of liquid detergent. A few decades ago, these same aisles would have been lined with cardboard boxes, drums, or plastic containers. The boxes and drums have been relegated to the end of the row, squeezed into a much smaller area.

But don't assume that powder detergents are a relic of the past. They're still the number-one choice across Africa, India, China, and Latin America, not to mention the most widely distributed form of detergent in many other countries around the world. They still have a role to play at home as well, so let's not be deceived into thinking liquid detergents are simply the water-based

> **WARNING!** When a bottle of detergent is almost empty, some people water it down so it lasts a bit longer. I don't recommend this for detergent or for dish soap, hand soap, shampoo, or basically any home cleaning product. I say this for two reasons: Firstly, adding water alters the solution's pH and overall composition, potentially impairing its performance. Secondly, and most importantly, diluting what's left of the product also dilutes the preservatives, rendering them potentially ineffective and leaving the liquid vulnerable to bacteria and mold. If you don't want to waste that trickle at the bottom of the bottle, add just enough water to get it out, use it right away, then toss the container into the recycling bin. Or, if you buy detergent refills like me, just top up the bottle when it's running low.

version of powder ones. The ingredients in each are very different, with different advantages and disadvantages.

Both categories offer a range of quality, from basic get-the-job-done varieties to high-end options with premium characteristics. It's up to you to decide which one fits your purpose—just don't fall into the trap of thinking they're all equal. Even if the differences in price are not always dictated solely by raw material costs, they do give some idea of the differences in function. Clearly, it's not easy to get an objective picture when you don't have the time or desire to wade through the ingredient lists on each manufacturer's website—which is why I decided to write this book!

POWDER

The simplest powder detergent may contain only a single anionic surfactant and a limescale inhibitor. It will be fairly inexpensive, and in many developing countries, this kind of basic product is often used for handwashing. At the other end of the spectrum, there's the range of machine-wash powder detergents marketed in more developed countries, which contain multiple surfactants, limescale inhibitors like zeolite, bleaching agents, stain-removing polymers, optical brighteners, fragrances, enzymes, and any other new developments the industry has come up with.

There was a time when consumers preferred large boxes of cheaply priced detergent. When I was growing up in Italy, I remember manufacturers going head to head on television to see who had the biggest box of laundry powder on offer. However, the increase in box size did not always equate with an increase in the quantity of powder inside, especially if the contents had been

specially formulated at a lower density (bulk density is the amount of powder by weight in a particular volume). There are ways of producing detergent granules that have a lower density (or inversely, that occupy a larger volume at the same weight). Another tactic was adding more of the less-expensive limescale inhibitors while keeping the amount of more-expensive surfactants the same. You'll often see bulk density tricks used in in processed ice cream. Have a look the next time you're at the store: Two tubs of equal size can have different weights due to the different amounts of air incorporated into the product.

Consumers have changed, though, and the enormous drums of powder detergent that once graced our aisles have disappeared. Super-concentrated alternatives rolled in around 1989, winning people over with their compactness, which allowed for smaller doses in the washing machine and less space taken up in the shopping cart and at home—although it took a while to convince those who saw it as a choice between very differently sized products at alarmingly similar prices.

Powder detergents often contain a stain-busting bleaching agent (usually sodium percarbonate), meaning that unlike for liquid detergents, there's no need to add one separately.

One disadvantage of powder detergents is how slow they are to dissolve at low temperatures. 140°F (or 60°C) washes used to be the norm, but with rising costs and increasing pressure to save energy, water temperatures of 85 to 105°F (or 30 to 40°C) have become more common. The trend is toward cold washing, which some countries have already embraced. This is a problem for powder detergent: It dissolves painfully slowly in cold water and can also leave behind residue.

Another issue is that they tend to absorb moisture from the surrounding environment, causing the powder to clump. Said clumps then have to be broken up, often scattering powder everywhere.

LIQUIDS

Liquid detergents were introduced as soon as 1956 in the US but only began to win serious market shares in Europe starting in the early 1980s. Nowadays, they are most Western consumers' preferred option.

These viscous liquids offer a host of advantages over their powder predecessors: They are easier to use, dissolve quickly and fully at 85°F (or 30°C), don't scatter powder everywhere when you're loading the machine, and come in containers that can be easily refilled.

Liquid detergents contain a broad spectrum of surfactants—far more than the powder versions—including a selection of nonionic surfactants that act on oily dirt. Since these nonionic surfactants are liquids, it's impossible to include them in powder formulations.

The limescale inhibitors of choice in powder detergents tend to be zeolite and sodium carbonate. Zeolite is an insoluble solid that's very difficult to suspend in liquid detergents, which tend to feature alternative substances likes sodium citrate or polycarboxylate. Polymers, which prevent dirt from redepositing on fabric, are also more difficult to incorporate into liquid detergents.

One advantage of liquid detergents is that they can be poured onto the fabric itself to pretreat stains before washing. As liquids, they dissolve fully in water during both hand- and machine-washing. The latter has become increasingly important as wash temperatures have dropped.

The biggest drawback of liquid products is that they contain no bleaching agent. The two don't function well together—the bleach would oxidize all the other ingredients before the bottle made it anywhere near the shelf. This is why it is recommended to add a mild bleach or equivalent product into the appropriate compartment in the detergent drawer when using a liquid detergent. It also explains why liquid detergents never took off initially (in Europe at least) when they were originally launched in 1981. Consumers lamented that they weren't as effective as powder detergents, and rightly so: Many of the powder formulations they were accustomed to had contained bleaching agents for decades, meaning any consumers who tried the new liquid variant on its own found the results extremely disappointing.

Even adding enzymes to a liquid detergent can be tricky, since proteases—which are great for removing protein-based dirt—end up destroying each other as they are proteins themselves. The same goes for other enzymes. Therefore, manufacturers have had to come up with ways of preventing this—for example, introducing techniques like microencapsulation to coat the enzymes and protect them from the detergent's other components.

TABLETS

The two traditional forms of detergent—liquid or powder—now come in a wide range of convenient-to-use options. Powders can be premeasured and compressed into tablets, which is popular for dishwashing detergents but less so for laundry detergents. Tablets are much more practical than their powder predecessors, as they don't spill or make a mess, you can easily gauge how many are left with a quick glance in the box, and they take up less space due to their compressed form. On the flip side, the fact that they come premeasured means you can't adjust the amount for more heavily soiled loads or for hard water. To work effectively, tablets must dissolve quickly and fully, or they will clog up the detergent dispenser. In terms of performance, they clean just as well as traditional powders.

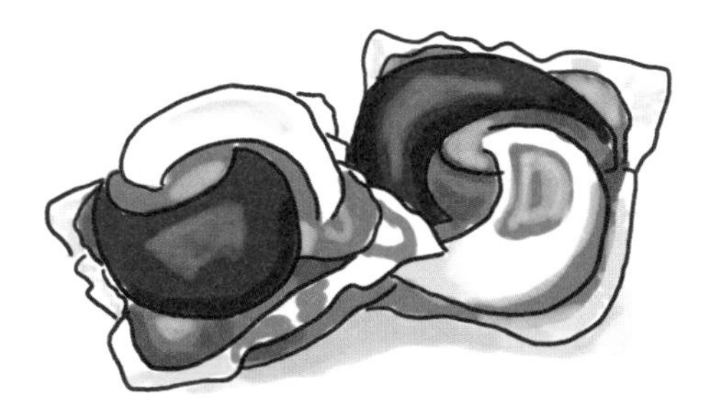

CAPSULES

Grocery stores also offer liquid detergents in small, single-use pouches, better known as capsules or pods. Just like tablets, they come with the restriction that someone else has set the dosage for average dirt and water hardness, but their advantage is that they're extremely convenient. Manufacturers have obviously had to overcome a few key issues, such as containing a water-based solution inside a film that must also dissolve in water. But consumers don't need to worry about that—all we need to know is that pods are as effective as a standard liquid laundry detergent.

Changing times

A lot has changed over the years: the kinds of dirt we need to wash, the frequency at which we wash, and even the type of clothes we are washing. It should come as no surprise, then, that the detergent market itself has undergone an equal number of changes.[16] Work clothes used to get far more soiled, people did more manual labor, more food was cooked at home, and more meals were eaten around tables covered by large linen tablecloths. The standard attire of an office worker, or what my dad referred to as your "Sunday best," was a white, or perhaps light blue, cotton shirt. People didn't take as many showers or baths. Laundry was done roughly once a week, and it was a long, drawn-out event. No matter what it took, clothes had to emerge cleaner than clean. To convince consumers that their product would produce the best result, laundry titans would often go head-to-head in a "my product cleans better than yours" advertising battle.

Things are different nowadays. We wash ourselves and our belongings more frequently, and our clothes get less dirty. We own far fewer white cotton shirts and wear far more dyed fabrics, the colors of which we try to protect for as long as possible. We wash at much lower temperatures to keep our electricity bills down and to protect the environment.

Nevertheless, solid detergents continue to work the best because liquid detergents are, by their nature, weaker. In 2018, the German consumer organization Stiftung Warentest, which investigates and compares goods and services, tested twenty-six different detergents on white clothes. The five liquid detergents included in the sample finished in the bottom five spots! This is why washing powders still have a role to play in Germany and around the developed world. Admittedly, powders are up against the growing market power of liquid detergents that promise to

16 I realize that many of my readers will be far too young to remember most of these changes.

COMPARING POWDER AND LIQUID DETERGENTS

POWDER	LIQUID
Better on non-oily dirt	Better on oily dirt
Cheaper	More expensive
Fewer surfactants	More surfactants
More water softeners (sodium carbonate and zeolite)	Fewer water softeners (sodium citrate)
Bleaching agent (sodium percarbonate)	Very few or no bleaching agents
More enzymes	Fewer or no enzymes
Includes aids to help clean at low temperatures	Can clean at low temperatures without aids
More effective at 140 to 195°F (or 60 to 90°C)	More effective at 85 to 105°F (or 30 to 40°C)

care for our colors, be gentler on our delicates, and remove body oils more effectively.

Meanwhile, in other parts of the world, more and more people are buying washing machines and giving up washing by hand. Other cleaning preferences vary by location, too: In North Africa and the Middle East, light-colored clothes are popular, making powder detergents with their associated whitening properties the preferred choice. Asian consumers have been found to be particularly attuned to the more powerful destaining, deodorizing, and whitening properties of ingredients like sodium percarbonate.

DID YOU KNOW?

People wash at much lower temperatures in the United States than in Europe, often choosing cold cycles. Therefore, detergents should be less effective at whitening in the US (although in reality, the ingredients often vary from place to place, so you can't tell the difference).

Hand- or machine-wash?

Despite the trend I just mentioned, many people around the world still wash by hand, and there's still a market everywhere for products that help with this, especially when it comes to laundering wool and similar fabrics. People prefer to hand-wash delicate items to save them from being spun around mechanically in a drum. Some washing machine settings even offer a hand-washing equivalent that uses more water and involves only the gentlest spinning.

In terms of lathering, consumers have come to believe that suds are the sign of an effective wash. This probably stems from the days when using a bar of soap was the standard. (Although, to be strictly accurate, soap lathers very little in hard water and is not the most effective of cleaners because of the way it bonds with calcium and precipitates.)

Things have changed since the bar-soap days, and now we have detergents. Nevertheless, the association between cleaning power and soap suds has persisted, with the result that consumers continue to expect a healthy amount of sufficiently dense lather. To achieve this, manufacturers of hand-wash products have added a suitable surfactant: Anionic surfactants provide great lathering. However, they are the worst for skin, so they can only be used in limited amounts. Protease enzymes also cause irritation, since protein in dirt and protein in our skin are all the same to them.

Wool and silk

Like other fibers obtained from animals, wool and silk are made of proteins. Since proteins are broken down by substances with basic pH values (or very acidic ones), the pH values of wool detergents must be neutral or only mildly acidic. If you don't want your favorite sweater to end up felted (see the box on the next page), steer clear of normal detergents containing alkaline ingredients (or soap, which is even more unsuitable). Wool and silk detergents contain mostly nonionic and cationic surfactants, as their positive or negative charges act like fabric softener, keeping the wool fibers light and soft. In a similar vein, proteases are an absolute no-no, as they will destroy your favorite silk shirt.

Wool felting and shrinkage

Wool fibers are formed mainly from protein—keratin to be more precise—and feature multiple bonds called disulfide bridges that connect sulfur atoms. Interaction with a base breaks these bonds and irreversibly changes the shape of the protein. Wool fibers have the same special structure as our hair—imagine that wool fibers and hair strands are covered in scales. An alkaline pH causes these scales to rise, which is why washing your hair with soap is not a good idea.

The mechanical action of a washing machine causes wool fibers to completely bind together, which we call "felting." Indeed, felt is the compressed fabric obtained from felting either wool or animal hair (from camels or rabbits, for example) in a series of steps that involve soap, hot water, and friction.

Hot water alone will not felt wool, but it will cause it to shrink. The wool's proteins denature and the resulting change in shape draws the fibers closer together. When this happens, there's still time to partially avert any permanent damage by washing the item again and spending some time stretching and pulling it back into shape. Be warned, though, that the fabric will never be as soft as it was originally, since protein denaturation is an irreversible process.

HOW WOOL GETS FELTED

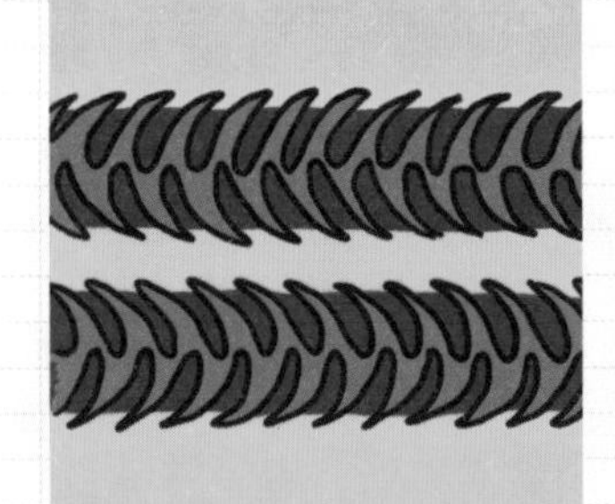

Wool fibers contain proteins with bonds connecting sulfur atoms

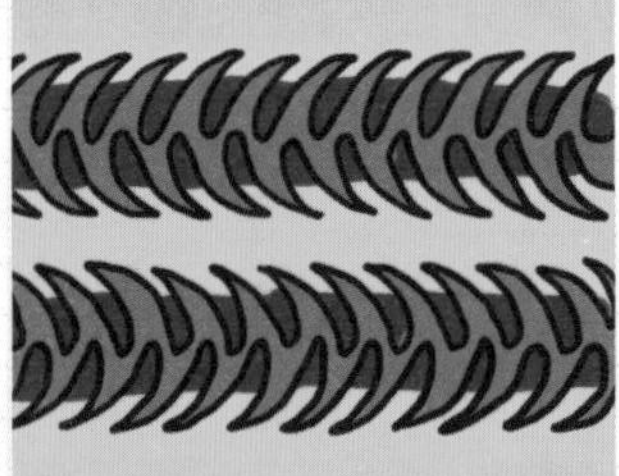

When a base is added, these bonds break, changing the proteins' shape

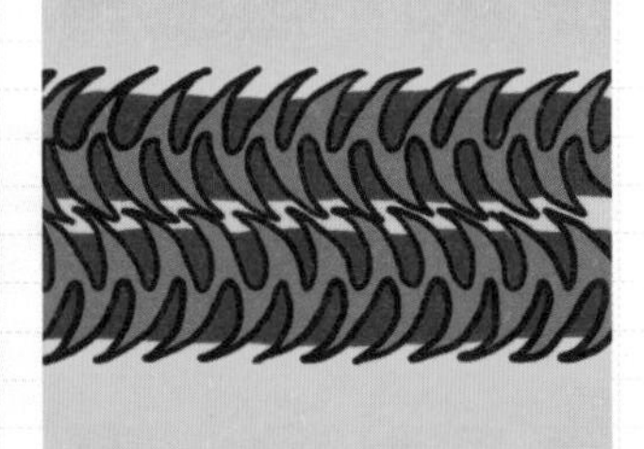

Mechanical action binds the misshapen fibers together

6

Chlorine-Based Bleach

The detergents we use to clean our clothes (and other items) are formulated to remove all kinds of dirt from all kinds of materials in all kinds of ways. We've seen how well they handle oil and grease, protein residue, and microscopic particles of both solid and liquid matter—but there are other types of dirt, like coffee or pasta-sauce stains, that ordinary detergents don't cope with very well.

Enter another class of cleaning product: bleaching agents. Most people will have either a bottle of bleach or a whitener like hydrogen peroxide at home in a cupboard somewhere. Both substances can be used as a standalone solution or mixed with another cleaning product, sometimes as part of a ready-made detergent formulation. Chemists have identified an array of different bleaching agents over the past couple of decades, but in the domestic cleaning market, two categories prevail: chlorine-based cleaners like bleach, which we'll discuss in this chapter, and oxygen-based cleaners like hydrogen peroxide or sodium percarbonate, which are handled in the next chapter.

Chlorine- and oxygen-based bleaching agents have long been used alongside detergents, primarily for removing stubborn fabric stains that those detergents fail to budge. The role these bleaching agents play in the laundry process varies from place to place. People in some countries, like the United States, add bleach at the same time as laundry detergent, while those in other countries, like Italy, tend to use bleach first, either as an entirely separate pretreatment or as part of a prewash cycle.

DID YOU KNOW?

Bleach doesn't clean: It whitens, but it doesn't remove dirt. Although it may make stains invisible and can start breaking down dirt, it doesn't take that dirt away. A surfactant added to a cleaning product will finish the job after bleach performs its function.

"To bleach" means to make a material whiter. This definition doesn't explain how the whitening occurs, so the word "bleach" could be used to refer to any chemical product capable of removing color from textile fibers (either from the whole material or just at the site of a stain). That said, "bleach" is commonly understood to mean a mixture of water and sodium hypochlorite, which is how the word will be used in this chapter.

When a whitening product does not contain the element chlorine, its name tends to be prefaced with terms like "delicate" or "color safe." In this case, the whitening effect comes from a completely different set of molecules called peroxides, not from sodium hypochlorite.

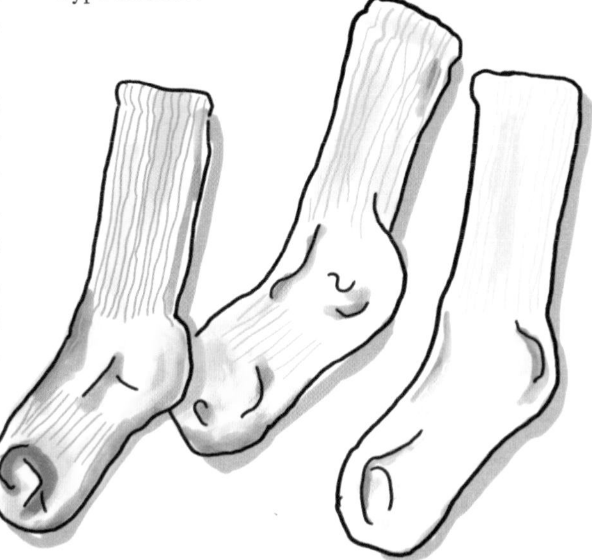

THE HISTORY OF CHLORINE-BASED BLEACH

In 1785, French chemist Claude Louis Berthollet was the first to recognize that percolating chlorine gas through water produced a liquid that could whiten fabric. His method was soon adopted by a number of textile manufacturers across Scotland, Belgium, and France, but the hazardous and highly irritating nature of chlorine gas harmed workers, damaging their lungs and leaving them with serious health problems. The chlorine also damaged the very textiles they were trying to whiten.

Another step forward occurred in the late eighteenth century, when the first hypochlorite-based commercial bleach, named "eau de Javel" after the Parisian borough where it was manufactured, was sold in France. The idea once again came from Berthollet, who perfected a process in which chlorine gas was passed through a weak solution of potassium hydroxide to produce potassium hypochlorite. Fast-forward to 1820, when pharmacist Antoine Germain Labarraque suggested replacing potassium hydroxide with lye as a cheaper option. This produced sodium hypochlorite, which rapidly supplanted potassium hypochlorite in commercial bleaches and disinfectants.

During and after World War I, the paper industry also drew heavily on the whitening ability of hypochlorite, as the routine manufacturing and transportation of cylinders of chlorine gas (also used as a chemical weapon) and concentrated lye made this substance more accessible. Large containers of liquid bleach were also first sold for domestic use around this time.

DID YOU KNOW?

Bleaching agents are highly reactive substances (they wouldn't be as effective if they weren't!), which means they can whiten your clothes and kill off any unwanted microbes at the same time. This is why they also work well as disinfectants, keeping hard surfaces around your home—especially in the bathroom and kitchen—clean and germ-free.

Chlorine

Swedish chemist Carl Wilhelm Scheele discovered chlorine gas in 1774. He observed that when he bubbled the gas through water, he produced "chlorinated water": a weak acid formed from hydrochloric and hypochlorous acids.[1] Scheele realized that this acidic water had the ability to remove the color from some vegetable pigments.

1 The equation for the formation of this chlorinated water is $Cl_2 + H2O \rightleftharpoons HClO + HCl$.

HOW BLEACH WORKS

In the spirit of chemical pedantry, I should clarify that hypochlorite ions[2] don't actually drive bleach's whitening action. The whitening we see is actually thanks to molecules of hypochlorous acid,[3] which is almost a hundred times more effective than the hypochlorite ions it comes from. Hypochlorous acid is a very weak but highly oxidizing acid. When hypochlorous molecules oxidize other molecules, breaking down the chemical bonds responsible for their colored appearance, they also destroy themselves. While some hypochlorous acid disappears in this reaction, more is made by the reaction of the hypochlorite ions with water (also known as hydrolysis).

Since a hypochlorite solution is a strong oxidizer (that is, it can change the structure of other molecules), it almost always removes the color from clothes. After all, that's why we use it! However, this strength also means that hypochlorite destroys both a red ink stain that you want to remove from a white T-shirt and a red T-shirt that you want to stay red. For this reason, you should use bleach only on clothing that's white and preferably made from cotton or polyester (a synthetic fiber that is not damaged by hypochlorite).

Garments with particularly stubborn stains can be steeped in bleach prior to washing, although you shouldn't do this for too long—definitely not until the bleach solution dries out. After pretreating an item with bleach, you should usually pop it straight into the machine for a wash cycle.

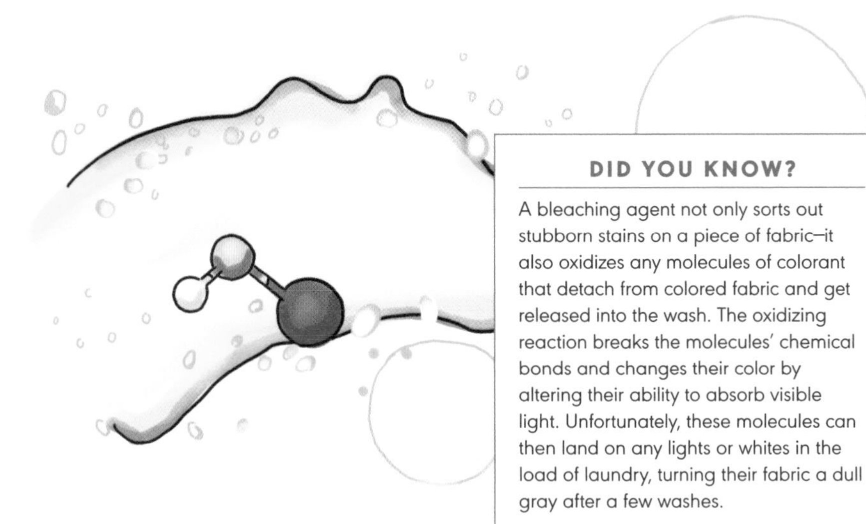

DID YOU KNOW?

A bleaching agent not only sorts out stubborn stains on a piece of fabric—it also oxidizes any molecules of colorant that detach from colored fabric and get released into the wash. The oxidizing reaction breaks the molecules' chemical bonds and changes their color by altering their ability to absorb visible light. Unfortunately, these molecules can then land on any lights or whites in the load of laundry, turning their fabric a dull gray after a few washes.

2 The chemical formula for hypochlorite ions is ClO^-.

3 The chemical formula for hypochlorous acid is $HClO$.

KEEPING BLEACH EFFECTIVE

Most of us are aware that medicine and cosmetics have expiration dates, but we usually don't notice that the same goes for detergents and whiteners. If it's stored incorrectly, bleach degrades faster than most other cleaning products, and expired bleach can easily disappoint you when it comes to getting that nasty stain out of your favorite white shirt.

When sodium hypochlorite spontaneously decomposes, it permanently loses its bleaching ability.[4] We can't stop this, but we can slow it down. Some factors may be out of our control, since they're determined by the manufacturer's formulation choices, but there are others we can do something about.

DID YOU KNOW?

Sodium hypochlorite breaks down molecules containing phosphorous so effectively that it is used as a decontaminating agent to remove traces of nerve gas during war.

pH

pH is the most important factor in bleach's stability. Above a pH of 11, decomposition is slow and the product stays relatively stable. At a lower pH, decomposition happens faster. The degradation liberates hydrochloric acid, which further lowers the pH and accelerates the process even more. To prevent this drop in hypochlorite concentration, commercial bleach formulations contain a small amount of lye to keep the pH stable at around 12 to 13.5.

A pH above 11 lessens the chlorine smell normally created by hypochlorous acid, since the charged hypochlorite ions that exist at this higher pH can't spread through the air as easily, and the proportion of hypochlorous acid present in the liquid is kept to a minimum.

When the pH drops too low, hypochlorous acid becomes predominant and the solution turns overly aggressive. At a pH below 7.5, which is close to neutral, bleach has the highest potential to damage the cotton cellulose fibers in our clothes. This is why Scheele's original chlorinated water was replaced with a pH-controlled hypochlorite solution as a bleaching agent: The pH of the former was too low and caused too much damage to textiles.

Metals

Metals also accelerate decomposition. For this reason, manufacturers have to be scrupulous when it comes to the purity of their raw materials: Even the tiniest trace of copper, nickel, or cobalt causes sodium hypochlorite to release oxygen, convert to sodium chloride, and lose its effectiveness as a bleach.[5] In this case, decomposition is minimal compared to the problems caused by pH, but a higher temperature makes metal-related issues worse.

Light

Light also accelerates the decomposition process by breaking down hypochlorous acid into hydrochloric acid, which is why bleach bottles are opaque.

---

4 Sodium hypochlorite decomposes to sodium chlorate and sodium chloride: $3NaOCl \rightleftharpoons NaClO_3 + 2NaCl$.

5 This is the reaction: $2NaClO \rightarrow 2NaCl + O_2$.

RELATIONSHIP BETWEEN BLEACH CONCENTRATION AND STORAGE TEMPERATURE

Concentration

The higher the concentration, the more rapidly the product decomposes. This is a problem for large-scale industrial production, since sodium hypochlorite is often transported at strengths of 12.5 percent and higher. The large bottles we buy for domestic use come in lower concentrations, usually around 5 percent, and have a much lower rate of decomposition. A 5-percent hypochlorite solution, stored for one year at 68°F (20°C) in an opaque bottle, will retain at least a 4.5-percent concentration of useful product.

Temperature

The higher the temperature, the faster the decomposition. Does this mean that dissolving bleach in hot water robs it of its whitening power (as claimed by a host of internet "cleaning gurus")? No, not necessarily, as you'll see in a moment. But the same bottle of bleach, stored for one year at 105°F (40°C), will lose more than half of its concentration.

Don't worry—the bleach we use at home typically encounters far less dramatic conditions. But let's consider what happens to this substance at very high temperatures. Generally speaking, bleach's speed of decomposition multiplies two- to four-fold with every increase in temperature of 18°F (10°C). Going back to our example of a 5-percent bleach, if it were heated to 176°F (80°C), its effectiveness would be cut in half in little more than a day—then halved again in another day, and so on. At 122°F (50°C), it would take more than ten days for the bleach to drop to less than half its original concentration. So if you use a little bleach in hot water for ten minutes, it's really not a big deal. Then again, you could just use cold water, as the bleach will be just as effective!

DID YOU KNOW?

If you see "aqua" listed as an ingredient in a product, don't worry—it's not a mistake. "Aqua" is the official name for water in the International Nomenclature of Cosmetic Ingredients (INCI), a standard that is also used for some cleaning products.

DAMAGE FROM BLEACH

Cotton fiber damage

Hypochlorite reacts with a wide range of natural substances, not just the colorants we want it to break down. Specifically, it has the potential to react with cellulose, meaning that, in the long term, cotton fibers in clothes can be damaged if bleach is used repeatedly. This kind of risk to cotton fibers increases the closer the pH gets to 7. When you pour bleach into water, the pH of the resulting solution depends on the quantity and hardness of the water, but it's typically somewhere from 8 to 9.5.

Have you ever had an old white garment literally fall apart in your hands? The fibers were probably weakened by being washed with bleach one time too many. If you want your whites to stay white *and* stay together, try mixing bleach with detergent when you steep the fabric. The alkalinity of the detergent helps to stabilize the hypochlorite and stop it from attacking the cotton fibers.

Color damage

The colorants used to dye fabrics usually have a different chemical structure than the colorants in the food that often ends up splashed on our shirts and tablecloths. They are similar enough, however, that they are both attacked and broken down by bleach. As an analogy, bleach is to stains as antibiotics are to bacteria: They both destroy without distinction, with bleach wiping out all color in the same way that antibiotics take out the good bacteria along with the bad.

Sadly, once hypochlorite has altered the molecules that give a piece of fabric its color, the damage cannot be reversed. If you get a splash of bleach on something colored and don't douse it in water right away, you'll end up with white "stains" that you can't dislodge—because there is nothing left to remove! You'll just have to accept the color is gone and the only way to get it back is to dye the garment all over again.

Pure sodium hypochlorite is a pale yellow solution, but the slightest contamination with iron turns it pink or red. So if you ever see an unexpected red

Creating sodium hypochlorite

Today, sodium hypochlorite is still made by reacting chlorine gas with lye (aka sodium hydroxide).[6] These raw materials are produced from the electrolysis of a sodium chloride solution (that is, an electric current passing through a solution of water and salt). The resulting hypochlorite has a concentration of 5 to 15 percent.

6 The reaction looks like this: $Cl_2 + 2NaOH \rightleftharpoons NaOCl + NaCl + H_2O$.

splotch in the basin where you're steeping some clothing or in the washing machine while you're running a bleach prewash, get your clothes out fast! The red marks are a sign that iron ions have oxidized and can get all over your shirts. Since bleach is also an oxidizer, it can't help with rust stains (but page 129 has some suggestions for fixing them). Note that iron isn't the only substance capable of leaving pink and red stains on your whites—for instance, the molecules in some sunscreens can stick to T-shirts or underwear and do the same thing (see page 133 for more on why).

WARNING! Never, ever, EVER mix bleach with another substance unless you are 100 percent sure what will happen. You should especially avoid combining bleach with acids.

Every year, hundreds of people end up in the hospital after intentionally mixing acidic cleaning products with bleach. When bleach comes in contact with an acid, it liberates poisonous chlorine gas, which was used as a chemical weapon in WWI due to its toxicity. Unfortunately, there are a host of toilet cleaners out there that are identical in everything but color, except that some contain bleach and some contain hydrochloric acid. While cleaning the house one busy morning, you could easily finish one bottle and start using another without thinking—but the new bottle happens to contain a hydrochloric acid-based descaler, and before you know it, you're choking on chlorine gas.

Separating bleach and acids also means avoiding using a toilet that you've just poured a bleach-based product into. Urine is acidic, so if it hits the bleach, don't be surprised when you get a whiff of gas. And don't forget that vinegar and lemon juice are acids, too.

Ammonia and bleach are both frequently used around the house. While they're equally effective cleaners on their own, when combined (which I beg you never to do), they can create a very unwelcome—not to mention highly irritating and toxic—substance called chloramine.

Bleach mixed with hydrogen peroxide (which is a weaker oxygen-based bleach) produces an instant whoosh of oxygen bubbles. On its own, oxygen is not toxic, but the fizz is so vigorous that it can easily send splashes of liquid onto your skin and eyes. Please also steer clear of bleach and ethyl alcohol mixtures, which can create a number of organic compounds—from chloroform to acetaldehyde—in varying concentrations.

So I'll say it again: Don't mix bleach with any cleaning product, really. Ever!

PROS AND CONS OF USING BLEACH

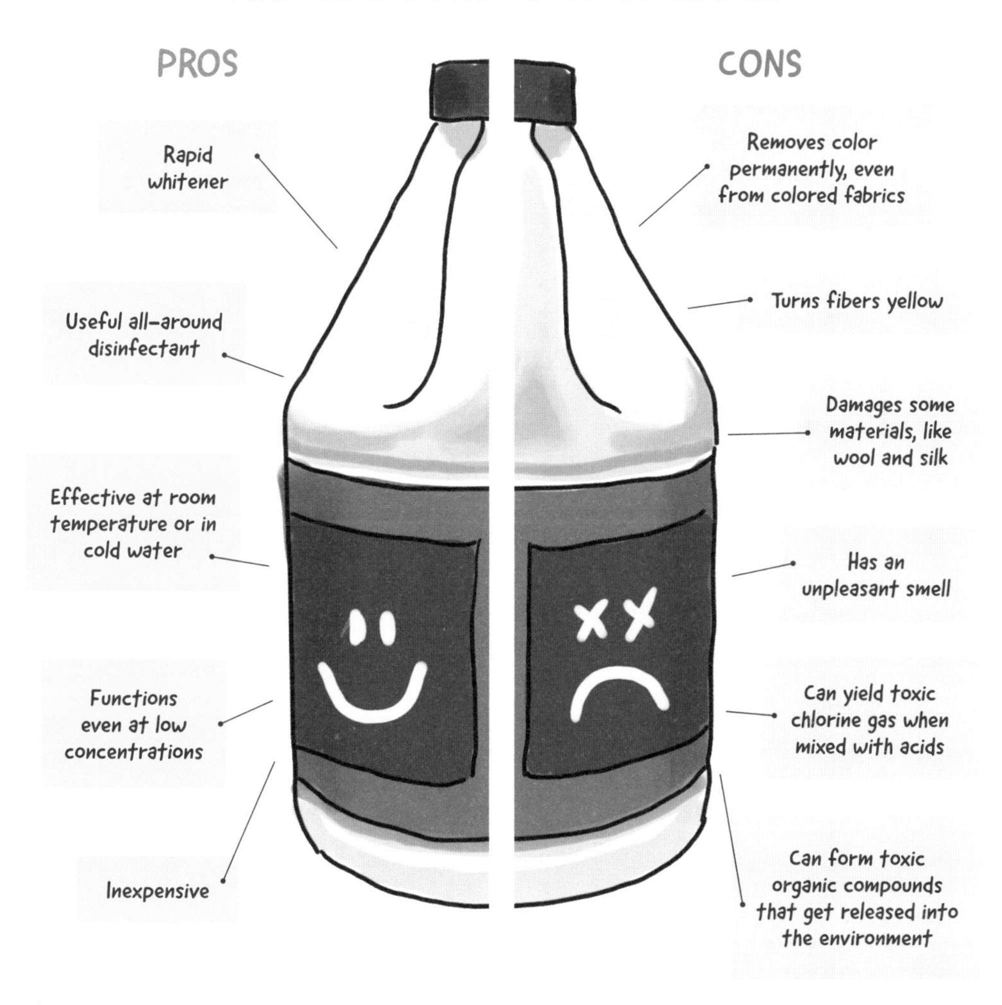

WHAT'S THAT SUBSTANCE?

Hypochlorous acid is weak, around seven hundred times weaker than the acetic acid in a bottle of vinegar. However, it's still an extremely aggressive and powerful oxidizer. It is important not to confuse an acid's strength and the pH it can reach in solution with its other properties, such as its ability to oxidize.

CLEANING DRAINS WITH CHLORINE

Sodium hypochlorite attacks fats, carbohydrates, and proteins, all of which are common drain blockers. The drain-cleaning products you're likely to find in your grocery store are made either with sodium hydroxide (also known as lye or caustic soda) plus a highly basic substance, or with a strong acid like sulfuric acid. In recent years, many high-concentration sodium hypochlorite and sodium hydroxide products have made it onto the shelves. These products are dense and viscous enough that you can pour them down a stopped-up drain even when it's filled with water.

DID YOU KNOW?

Sodium hypochlorite is often added to the water supply to make sure the water coming out of our faucets is safe to drink and free of harmful microbes. Sometimes the odor can be a little unpleasant depending on how much hypochlorite has been added. There's no need to worry, as the water is still perfectly drinkable—but if you'd rather it didn't smell so strong, then pour yourself a cup of water and let it sit for a few minutes. The hypochlorite will convert to chlorine and evaporate.[7]

CHEMISTRY CORNER

Available chlorine

Labels on chlorine-based products often state their percentage of "available chlorine" (called $AvCl_2$) or "active chlorine," even though they don't contain any chlorine gas and their whitening ability doesn't come from this substance. This information allows consumers to compare the oxidizing ability of chlorine-based products with varying compositions, since a product's oxidizing power derives entirely from the presence of chlorine. The percentage weight of hypochlorite in a solution is calculated by multiplying the available chlorine by 1.05. For our domestic purposes, it really makes no difference which number we use: percentage of hypochlorite or available chlorine.

7 The equation for this reaction is $HClO + Cl^- + H+ \rightleftharpoons Cl_2 + H_2O$.

WARNING! Sodium hypochlorite can cause severe irritation to skin and eyes. The level of damage depends on its pH, concentration, and viscosity as well as your length of exposure. Domestic bleach has a pH between 11 and 13, which isn't high enough to result in lasting issues if you treat the exposed body part right away. So if you accidentally spill bleach on yourself, remove your clothing immediately and rinse the affected skin under running water for at least ten minutes. If you are unfortunate enough to get a splash of bleach in your eye (did you forget to put your safety glasses on?), rinse your eye under running water for around ten minutes, being sure to keep it open.

Typically, the concentration and pH of the average household cleaner are not high enough to cause serious issues. Unfortunately, higher concentrations are less safe. There have also been reports of rare cases of hypersensitivity to sodium hypochlorite, which is known to cause allergic reactions that can trigger dermatitis, even in weak solutions.

AT THE GROCERY STORE

Bleach with extra features

We've learned that bleach is not a stain remover because it doesn't technically lift dirt. Its action is twofold: Firstly, it turns dirt colorless and therefore invisible, and secondly, it makes it easier for a detergent to pull away molecules of dirt that were previously firmly anchored to a surface. To fulfill consumer expectations, some manufacturers combine bleach with surfactants in order to offer detergents that can also whiten and disinfect. Such hybrid products are often sold specifically for use on bathroom surfaces. They typically come in gel form, so they can be used on vertical surfaces like toilets and bidets or delicately dabbed onto individual stains (which is much harder to do with liquid bleach). Further new entries to the market include combinations of bleach, surfactants, and abrasive particles (calcite or silica powder) that help remove dirt with gentle scrubbing.

I think I'm right when I say that not many people enjoy the smell of bleach. Manufacturers know this, too, and they have come up with formulations that try to offset the characteristic odor that wafts from a bucket of bleach solution or hangs in the air when a room's surfaces have just been wiped down. Fragrances do not alter bleach's ability to whiten and disinfect, although it's quite a chemical challenge to find a substance that can cover the smell of hypochlorous acid while remaining stable at such a high pH and in the face of such a powerful oxidizer. Generally, the molecules responsible for the fragrance are made into an emulsion so that they don't truly mix with the liquid solution.

SOLID BLEACH

Chlorine-based bleaches and whiteners also come in solid form. The first solid version was made from calcium hypochlorite[8] and introduced in Glasgow in 1798 by Scottish chemist Charles Tennant, who created it by reacting chlorine gas with slaked lime (aka calcium hydroxide). It was sold as bleaching powder and contained calcium hypochlorite, calcium chloride, and calcium hydroxide. This type of powder has a very high percent of available chlorine: up to 40 percent of its weight. Some stores still stock it. Before liquid bleach was introduced for consumer use, dissolving calcium hypochlorite powder in water was the preferred way of making bleach at home for at least a century.

The high pH values yielded by calcium oxide still make this type of solid bleach ideal for use in tropical areas or for water sanitization in developing countries. A more concentrated version containing up to 65 percent calcium hypochlorite is used to disinfect swimming pool water. However, it can't help with laundry due to its high calcium content. One other industrial-grade chlorine-based whitener and disinfectant is chlorine dioxide (ClO_2).

Hypochlorite stability

The graph below shows how the percentages of dissolved chlorine (Cl_2), hypochlorous acid (HClO), and hypochlorite ions (ClO^-) vary depending on a solution's pH. At a pH of 7.5, for example, half of the chlorine is present as hypochlorous acid and the other half as hypochlorite ions.

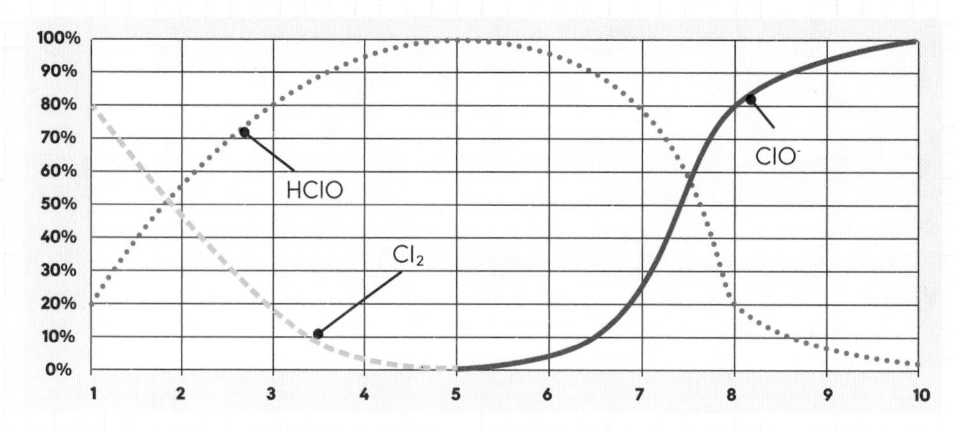

8 The formula for calcium hypochlorite is $Ca(ClO)_2$.

CLOTHING LABELS

I have to confess, there's something I should do but often don't. It has to do with those labels inside our clothes—yes, those annoying little ones, often made up of several smaller labels all stitched together, that make you itch behind your neck or somewhere around your waist. I don't know why they can't just be located somewhere more comfortable that doesn't rub every time I move. When it becomes too much, I unpick the stitches holding the labels in place with the help of a small pair of sharp scissors. (You have to be careful not to accidentally snip the fabric you're removing them from, of course. And cutting the labels straight off rather than undoing the stitches isn't an option, since it leaves a sharp edge that can be even more annoying.)

I know I'm not alone in doing this, but I also know it's wrong. Or rather, I know that I should memorize all the useful information on the labels before removing them. I'm talking about things like fabric composition and washing instructions found in those little hard-to-decipher symbols—triangles, circles, squares, and whatnot—that are all essential for looking after our clothes. These markings are often universal, but their exact use and appearance can vary regionally.

The symbol associated with bleach is the triangle. An empty triangle means the garment can be safely bleached with any kind of product. Any other additions inside or across the triangle signify what type of bleach to use and what type to avoid.

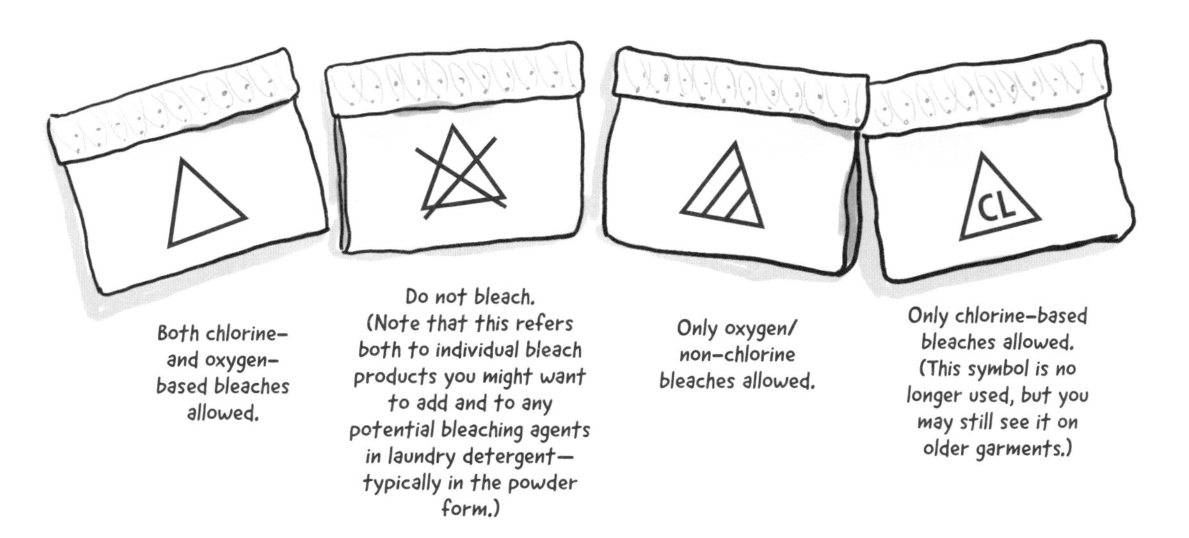

Both chlorine- and oxygen-based bleaches allowed.

Do not bleach. (Note that this refers both to individual bleach products you might want to add and to any potential bleaching agents in laundry detergent—typically in the powder form.)

Only oxygen/non-chlorine bleaches allowed.

Only chlorine-based bleaches allowed. (This symbol is no longer used, but you may still see it on older garments.)

BLEACH, THE ENVIRONMENT, AND YOU

When you consider how corrosive bleach is, how it wipes out anything bacterial that crosses its path, how it's highly basic, how it can harm you if it splashes on or near you, and how it can even create a poisonous gas . . . it would be understandable if you assumed that all chemical cleaners containing bleach present a major environmental hazard when they go down the drain. But the truth is, it's not at as harmful as it seems: As long as you use reasonable amounts and stick to the manufacturer's instructions, bleach's environmental risk is minimal.

As I mentioned before (page 97), hypochlorite is extremely reactive with organic substances, so when household hypochlorite reaches the sewage system, it is quickly destroyed. The nitrogen-rich substances in wastewater react instantly with hypochlorite, converting it to harmless sodium chloride. One study showed that more than 96 percent of hypochlorite poured down household drains is destroyed within two minutes, and others had similar findings.[9] Hypochlorite is long gone by the time wastewater reaches the sewage plant, which is good news—otherwise, there could be risks to the essential microorganisms used in water treatment.

Maybe the wastewater from your home feeds into a septic tank, where it is treated with bacteria before it reaches the main drainage system. There's really nothing to worry about here, either, given the small amounts of bleach we're talking about. And really, it makes no sense to throw buckets of bleach down the toilet, since very little is required for disinfecting—that old adage "the more the merrier" certainly doesn't apply here. Nevertheless, always read the label's instructions so you can be kind to the bacteria in your tank.

Based on all this, environmental safety agencies do not classify sodium hypochlorite as a hazard despite how toxic it is to aquatic life. In 1991, the Environmental Protection Agency (EPA) renewed its approval for drain disposal of bleach, concluding that registered uses of hypochlorite do not result in unreasonable adverse effects on the environment.[10]

Industrial-strength bleach is a different ball game, and so are situations in which hypochlorite and similar compounds are used in much larger amounts. In the final stages of wastewater treatment, for example, sodium hypochlorite is added at much higher concentrations than we would normally deal with at home, with the goal of killing off any remaining bacteria before the treated sewage is released back into surface waters.

To prevent the spread of viral and bacterial diseases, effluent must be thoroughly disinfected before it is released into waterways and reservoirs to eliminate any potential

9 W. L. Smith, "Human and Environmental Safety of Hypochlorite," in *Proceedings of the 3rd World Conference on Detergents: Global Perspectives*, ed. Arno Cahn (Champaign, Ill.: AOCS Press, 1994): 183–192

10 United States Environmental Protection Agency, *R.E.D. Facts: Sodium and Calcium Hypochlorite Salts* (Washington, DC: EPA, 1991).

pathogenic microorganisms that may have survived the water treatment process. Anywhere you go in the world, sodium hypochlorite is the chemical of choice for this kind of industrial process. However, that choice also brings with it a number of concerns. Firstly, the available chlorine can form chlorinated organic compounds that are potentially toxic to humans and must not be discharged into waterways. Secondly, any residual chlorine in the water has to be reduced or eliminated (using sodium bisulfite) to prevent any harm to flora and fauna. Finally, hypochlorite is usually the final step in the water treatment process, and if the water has not been sufficiently purified before it's added, chloramines can form, which can also damage aquatic life.

Overall, international environmental agencies apply different standards to large-scale or industrial uses of hypochlorite—such as for pest control, in the paper and textile industries, or by dry cleaners. Many countries also impose strict limits on the presence of chlorinated organic compounds in the water that leaves treatment plants.

Sodium hypochlorite can also react with household cleaners to create VOCs (volatile organic compounds), which are hazardous to human health in large quantities. However, diluting a little bleach in water releases only a very small amount of these materials. When bleach is combined with other organic materials, like perfumes, the quantity of VOCs released is much higher. There is little evidence that VOCs found in indoor air after cleaning with bleach and water are of any concern. That said, it's always a good idea to keep the room, or the whole house, well-ventilated when using chlorine-based products.

TRUE OR FALSE?

Only hospital-grade products can truly disinfect, and bleach is not one of them.

FALSE Left to react for around ten minutes, even hypochlorite with a 0.1-percent concentration (much lower than store-bought bleach) disinfects. Of course, if bleach has exceeded its shelf life or deteriorated for some other reason, its actual hypochlorite concentration can drop considerably and make it less effective.

Bleach should never be used on food.

TRUE If you need to make sure your food is bacteria-free, you can find hypochlorite formulations developed specifically for this purpose. These products are made to be safe for expectant mothers or people with weakened immune systems. Ordinary bleach was not designed for this purpose, not even in diluted form, because it doesn't meet food safety standards.

The oxygen released during the bleaching process is what actually causes the whitening.

FALSE Bleaching action comes from hypochlorous acid in equilibrium with hypochlorite ions.

Diluting bleach in hot water makes it instantly ineffective.

FALSE It takes days for bleach to lose its power like this, even when it's heated to 176°F (80°C).

USES FOR CHLORINE-BASED PRODUCTS

DID YOU KNOW?

The human immune system generates a small number of hypochlorite ions as a natural response to infection with viruses or bacteria.

Q&A: Chlorine-Based Bleach

What's the difference between standard bleach and delicate, gentle, or color safe bleaches?

The two products are made from completely different substances, despite the descriptions suggesting they are the same thing with two different strengths. Common bleach contains sodium hypochlorite, which is highly oxidizing and a powerful whitener. Delicate, gentle, or color safe bleaches contain hydrogen peroxide, which is a weaker oxidizer that can be used on colored fabric.

Does bleach remove mold?

Yes. Sodium hypochlorite has disinfectant properties and attacks mold when used at the right concentration for the right amount of time. However, your local grocery store has a variety of other mold and mildew removers that won't release chlorine gas around your house.

Why is the discoloration caused by bleach permanent, even when you try to redye the fabric the same color?

When a splash of undiluted bleach gets on a colored garment, you might not be able to restore it because the hypochlorite will probably also have damaged the textile fibers, leaving them unable to absorb any new color.

Why does bleach feel slimy and soapy when I touch it?

Bleach normally has an alkaline pH between 11 and 13, determined in part by the amount of residual lye it contains. When it comes into contact with the oils on your skin, it makes soap—which explains why it feels soapy!

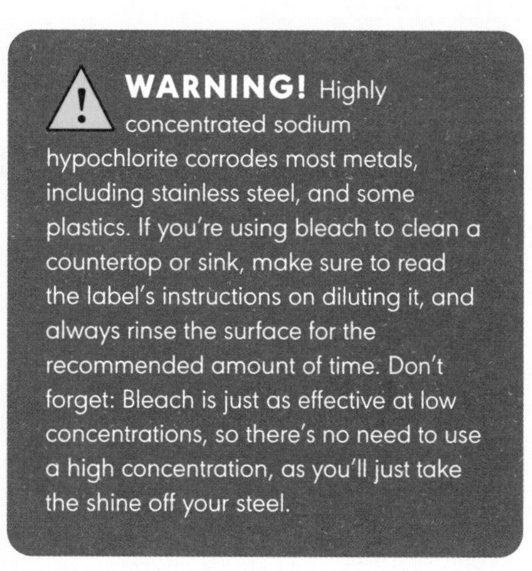

I accidentally put my bare hands into a bucket full of water and bleach. Now they smell like hypochlorite. What do I do?

Try dipping your hands into a 1-percent hydrogen peroxide solution, making sure to check the concentration of the hydrogen peroxide so you can dilute it properly. This will get rid of the remaining hypochlorite by turning it into oxygen and salt. Then, rinse your hands under the tap and apply some hand lotion, since both bleach and hydrogen peroxide can irritate your skin.

CAN I USE BLEACH . . .

To remove limescale?

No. It's completely useless for this purpose.

As a stain remover?

Yes, but only on white fabric that allows bleaching.

To wash clothes?

No. Bleach alone doesn't clean.

To disinfect?

No for skin or food, yes for hard surfaces, but make sure to follow the dilution instructions on the label.

To clean metal?

No. Many metals corrode quickly when exposed to bleach.

7

Oxygen-Based Bleach

Oxygen is the supreme symbol of life, both alone when it fills our lungs with air and together with hydrogen when it makes water. Without it, we couldn't survive. Yet in the ancient past, oxygen was behind one of the biggest mass extinctions in world history: Microbial species that didn't use it to live and thus weren't ready for the atmosphere's "great oxidation" were literally burned alive.

Two billion years later, we now breathe diluted oxygen effortlessly and have harnessed its oxidizing properties for cleaning purposes. When we go to the store, we can choose between chlorine- and oxygen-based bleaching agents. To be more precise, it's not actually oxygen itself, but rather hydrogen peroxide that does all the work in an oxygen-based product. As a whitener, hydrogen peroxide is not as strong as the hypochlorite in classic bleach, but this is not necessarily a bad thing. Oxygen-based bleach might be less effective at removing color from stains, but it also leaves the color of the garment itself relatively untouched, making it a much more versatile product than the chlorine-based version.

THE HISTORY OF OXYGEN-BASED BLEACH

Sodium perborate was the first chemical used as an oxygen whitener—it was originally added to Persil, a German brand of laundry detergent, in 1907. It took until the 1980s, however, for this substance to become more common. Then, advertising campaigns often drew attention to the amazing whitening powers it brought to detergents. Sodium perborate was most effective in 195°F (or 90°C) washes but was disappointing below 160°F (or 70°C) or in short cycles. This chemical was eventually removed from cleaning products when average wash temperatures dropped and one too many cases of harm to the human reproductive system were reported. After this, two types of oxygen-based whiteners rose to prominence: those containing hydrogen peroxide and those made with sodium percarbonate.

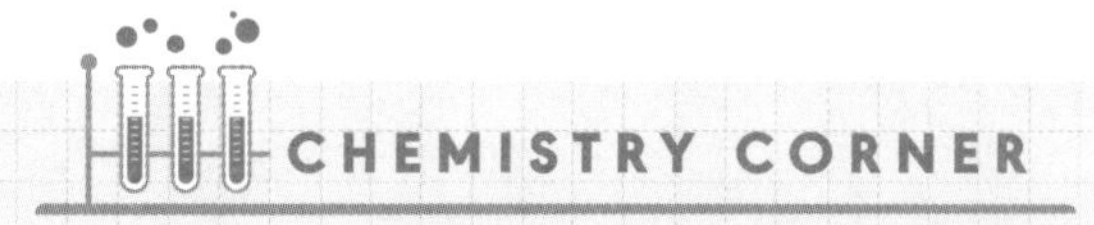

Active oxygen

The oxygen released from hydrogen peroxide, known as "active oxygen," was once believed to be the source of bleaching action, but recent studies have disproved this. In an alkaline environment, hydrogen peroxide dissociates into one hydrogen ion and one hydroperoxide (or perhydroxyl) ion, and the latter drives the whitening.[1]

1 The formula for this reaction is $H_2O_2 \rightarrow H^+ + HOO^-$.

HYDROGEN PEROXIDE

Hydrogen peroxide is a powerful oxidizer that removes a variety of different-colored stains. It has been present in our homes as a water-based solution for centuries, and we normally reach for it as a disinfectant or stain treatment. In its concentrated form, it's a pale blue liquid, but the dilute solutions sold for consumer use look almost identical to water. Most of the world's hydrogen peroxide is actually used in the paper industry for bleaching wood pulp.

The problem with hydrogen peroxide is that it's a weak acid, and in order for it to act quickly, the pH needs to be over 10.5 and the temperature above 122°F (50°C). Any less, and hydrogen peroxide takes longer to work or requires additional help to activate at lower temperatures. For this reason, it is generally advised to leave particularly stubborn stains to steep overnight, because at standard room temperature, hydrogen peroxide takes a long time to have any effect. But be careful, as the chemical evaporates as it works, and the higher concentration of the remaining product can damage garment fibers.

Hydrogen peroxide was first discovered in 1818 by French chemist Louis Jacques Thénard. However, it only became more prevalent as a whitening agent in consumer cleaning products in recent years. This is because to create a product with a decent shelf life, manufacturers had to find a way of stabilizing the chemical and preventing it from decomposing to yield oxygen (and thus becoming useless). The higher the temperature, the quicker hydrogen peroxide decomposes. It took until the 1990s to formulate an acidic medium in which the peroxide can remain stable for large amounts of time.

Because of its bleaching properties, hydrogen peroxide is often referred to as "delicate," "color safe," or "gentle" bleach. This causes no end of confusion for consumers, who mistakenly think that the whitener they've bought is a less-concentrated version of classic chlorine bleach, when it's actually a different chemical altogether.

Delicate bleach normally contains 3 to 7 percent hydrogen peroxide. It can be used in two ways: applied to a stain for pretreatment prior to washing or added to the wash cycle with the detergent. Typical product formulations have acidic pH values and contain both surfactants and chelating agents to capture any metal ions. The acidic medium required to keep hydrogen peroxide stable for a lengthy period curtails its whitening power until the peroxide comes into contact with an alkaline detergent and its power is unleashed.

DID YOU KNOW?

When hydrogen peroxide is poured onto an open wound, it fizzes. This is the work of the enzyme catalase, which decomposes the peroxide. Catalase is found in a variety of bacteria and is often the reason hydrogen peroxide fails to whiten some surfaces as well as it should.

WARNING! The general rule is never to mix chemical cleaners if you don't know what might happen, and this applies to hydrogen peroxide, despite the common—and, unfortunately, inaccurate—belief that it is harmless. Specifically, hydrogen peroxide added to an organic compound like acetic acid yields peroxyacetic acid, which is a much stronger oxidizer. Bleach plus hydrogen peroxide creates oxygen gas, which won't harm you by itself, but the reaction is so violent that the bubbles and splashes it produces might!

DID YOU KNOW?

Concentrations of hydrogen peroxide above 70 percent can detonate when mixed with other substances, creating an explosion just as violent as TNT or nitroglycerine.

Tips for cleaning with hydrogen peroxide

REMOVING BLOODSTAINS

Hydrogen peroxide does better at removing bloodstains when the blood is still wet and you add cold, not hot, water. (Hot water denatures the proteins in the blood, making whitening more difficult.) I find this works well on handkerchiefs and . . . walls! Mosquitoes drive me absolutely crazy, and whenever I see one, I can't stop until I've found and killed it. I've gotten so good that I can even catch them in my fist in a single move. However, I do still miss sometimes, and if one ends up splattered on the wall, a quick intervention with some cotton wool dabbed in hydrogen peroxide soon gets rid of the stain.

BRIGHTENING UP LEGO BRICKS

Hydrogen peroxide and UV exposure can make the colors of grimy old LEGO bricks vibrant again. You can actually use this trick to restore any plastic items made of ABS (acrylonitrile butadiene styrene), like shoes or toys. This marvel of chemistry is based on the same principle used to treat wastewater or bleach cotton fibers without chlorine in order to be more eco-friendly. It's easy peasy: Just put the LEGO bricks in a glass bowl, pour in some 3-percent hydrogen peroxide (the typical variety sold at the store), cover the bowl with plastic wrap, and leave it in direct sunlight for a few days. The sun's UV rays will break down the hydrogen peroxide to release free radicals,[2] which work their magic and attack the colored substances responsible for fading this favorite childhood toy.

I decided to try this with some old LEGO bricks I had in a box at home. To test the method's effectiveness, I split the pieces into

2 Chemists like me call them "hydroxyl radicals" (indicated by the symbol HO·). The dot beside the O represents the unpaired electron.

two groups and put half in a glass of hydrogen peroxide that I left in the dark, wrapped in a piece of tinfoil. Two days later, the ones that had been exposed to both hydrogen peroxide and UV rays were much cleaner than the ones I kept in the dark.

You can speed up the process with a UV lamp and some sodium carbonate (soda ash) to raise the pH of the hydrogen peroxide solution.[3] As always, make sure to wear gloves and safety glasses when working with chemicals.

DID YOU KNOW?

Hydrogen peroxide is used in some whitening toothpastes to remove stains from tooth enamel. It's also employed in the cosmetics industry as a bleaching agent for hair: Applied before dye, it removes hair's natural color by oxidizing the melanin and allowing a new color to take its place.

SODIUM PERCARBONATE

Sodium percarbonate comes from the reaction of hydrogen peroxide with sodium carbonate, an ingredient we've discussed several times already. Several decades ago, sodium percarbonate replaced sodium perborate as the preferred bleaching agent in powder detergents and stain-removing additives. In chemical terms, sodium percarbonate is a solid that includes hydrogen peroxide trapped in a sodium carbonate structure—and on contact with water, that hydrogen peroxide is released. Then, the remaining sodium carbonate (a base) aids in the cleaning process and accelerates the action of the hydrogen peroxide—which works much better in a basic environment than in the acidic medium of delicate bleach.

Sodium percarbonate is unstable in the presence of moisture, which is why it is added only to powder detergents, not liquid ones. Manufacturers using it as an ingredient have to prevent it from decomposing, which they do by encapsulating it in tiny spheres during production.

Percarbonate is a highly effective whitener at 140°F (or 60°C). Any hotter than this, and it rapidly decomposes, releasing the hydrogen peroxide too quickly. If 140°F is your preferred wash temperature for things like linens and sheets, a detergent containing sodium percarbonate will definitely do a good job.

However, many modern synthetic fibers don't wash well at high temperatures: The heat loosens the textile mesh, opening it up to dirt that can get trapped inside when the fabric dries. Also, the drive to be more energy-efficient has resulted in washing at lower temperatures—usually from 85°F (or 30°C) to 105°F (or 40°C)—and using shorter

3 Nanfang Wang et al. "An environmentally friendly bleaching process for cotton fabrics: mechanism and application of UV/H_2O_2 system." *Cellulose* 27, no. 2 (November 2019): 1071-83.

cycles, both of which are not ideal for sodium percarbonate. At these temperatures, it takes more than 24 hours for percarbonate to make any difference as a bleaching agent or disinfectant! For this reason, chemical substances called "bleach activators" are added to formulations developed for use below 140°F (or 60°C). They react with the hydrogen peroxide that sodium percarbonate yields, forming a more powerful oxidizer that drives the whitening: peracetic acid.

> **DID YOU KNOW?**
>
> Sodium percarbonate is environmentally friendly. If it's not activated and used, it decomposes into sodium carbonate and hydrogen peroxide, both of which pose no risk to the environment when used for household cleaning.

The first detergent to contain a whitener and activator was launched by Henkel in 1972. The activator in question was TAGU, an acronym for a very complicated molecule called tetraacetylglycoluril. The most widely used bleach activator in Europe nowadays is TAED (tetraacetylethylenediamine), which Unilever first added to the detergent Skip in 1978. SOBS and NOBS are the most widely used activators in the US and Japan—I'll spare you the chemical names.

These activators are what make sodium percarbonate (and perborate before it) effective at today's lower wash temperatures. Indeed, they're key ingredients in a number of powder detergents advertised as being tough on stains at low temperatures. To use TAED as an example, one TAED molecule yields two molecules of peracetic acid, another reason it has become the preferred activator for sodium percarbonate—two for the price of one! It works well at 105°F (or 40°C) and even better at 140°F (or 60°C). At temperatures higher than this, it is not as effective due to several complicated parallel chemical reactions that I won't go into here. At lower temperatures, it still activates the percarbonate, but since it's less soluble at lower temperatures, it needs more time to do the same work. At room temperature, it can be used to pretreat garments by leaving them to steep overnight.

> **AT THE GROCERY STORE**
>
> **Reducing bleaches**
>
> Most of the whitening agents in household cleaners are oxidizers. However, oxidation is not always necessary for bleaching—there are alternatives called reducing agents which, chemically speaking, are the opposite of oxidizing agents. Your local grocery store no doubt stocks these special laundry additives, which contain reducing agents like sodium dithionite ($Na_2S_2O_4$) or sodium bisulfite ($NaHSO_3$) and are just the ticket when colors have run in the wash and you need to remove unwanted dyes. Have a look the next time you're shopping—they might be able to fix something your ordinary bleach can't.

> **DID YOU KNOW?**
>
> 60 percent of the percarbonate produced worldwide is used in laundry powder, and Solvay is the leading global manufacturer. Lubrizol, a global supplier of specialty chemicals based in Ohio, is the world leader in TAED manufacturing.

WHEN BLEACH IS TOO EFFECTIVE

Current bleaching agents and activators are limited in their effectiveness by the amount that can be contained in a cleaning product. When they perform their functions, they react and get used up. It would be far better to have an activator that doesn't deplete itself in this way—what we chemists refer to as a "catalyst"—so we wouldn't need quite as much of it. Financial gain is the most obvious benefit, but it would also bring down the product's overall weight, requiring less energy for transportation and ultimately being more environmentally friendly.

Chemists set to work on this problem and came up with host of different catalysts. For example, in 1994, Unilever launched a range of products (Persil Power, OMO Power, and Skip Power) featuring a manganese-based catalyst. At 105°F (or 40°C) and with only 1 percent of the amount of TAED previously required, they did the exact same job. However, it wasn't long before trouble set in: A few months after the products were launched, consumers began to report that they were so strong, they were rotting clothes after only twelve washes. A new formulation was launched with 80 percent less catalyst, but it was still too powerful, fading colors and damaging fibers. By 1995, Unilever withdrew the products from the market, and chemists went back to the drawing board to find a catalyst that would not harm fabric.

CHLORINE- VERSUS OXYGEN-BASED BLEACHES

Chlorine-based	Oxygen-based
Better whitener	Less effective whitener
Causes more damage to fabric fibers	Gentler on fabric fibers
Fast-acting	Slow-acting
Can be used on whites only	Can be used on colors
Effective in cold water	More effective at 140°F (60°C)
More potentially hazardous	Less potentially hazardous
Higher risk of accidentally using too much	Lower risk of accidentally using too much

DEEP CLEANING WITH BLEACH

Using a bleaching agent, whether it's incorporated into a powder detergent or added alongside a liquid detergent, has a number of benefits related to removing unwanted bacteria and hygienically cleaning fabric.

I use the word "hygienic" intentionally, as I don't want to convey the idea that bleach can sterilize or disinfect when used in your home washing machine. And quite often, it wouldn't even make sense to try decontaminating your clothes. Of course, places like hospitals and nursing facilities have to prevent the spread of illness by carefully disinfecting clothing, bedding, and any other fabrics that might carry harmful germs. But in our homes, there's no evidence that we are any less likely to pick up infections if we go through the costly process of disinfecting everything we wash. I realize that some people have an extreme fear or dislike of pathogens and shiver at the thought of living in a world full of microorganisms (even if most are harmless and some are beneficial). Nevertheless, even if it were possible to disinfect the things we use around the house, bacteria and fungi would only start growing on them again the minute they come out of the washing machine.

However, you may still be interested in eliminating musty smells from freshly washed but still damp clothes—which are a breeding ground for mold and bacteria, especially if you don't have a dryer or a sunny backyard where you can hang them out to dry. Please note, though, that using a deep cleaner like bleach to rid your linens of germs before hanging them in a poorly ventilated indoor space defeats your purpose, as any fungi and bacteria in the room will quickly reestablish themselves on the clean laundry and leave it smelling musty.

Bleaching agents are actually useful for cleaning the washing machine itself and removing the sludge of germs, otherwise referred to as biofilm, that builds up inside and creates a constant source of contamination. This slimy substance makes your clothes smell bad and can damage the machine in the long run because of the corrosive substances that the microbes produce.

WHAT'S THAT SUBSTANCE?

PERACETIC ACID

Peracetic acid is produced by reacting acetic acid and hydrogen peroxide. It is a powerful disinfectant and oxidizer used in a number of detergents and as a method of disinfecting wastewater treatment plants. Peracetic acid is extremely fast-acting and is applied at very low concentrations.

Q&A: Oxygen-Based Bleach

Do I need to use color safe bleach even if my clothes aren't stained?

A little dye always escapes when you wash colored clothes, even if you don't notice this right away. Wash after wash, the colored molecules released from the fabric gradually mix with the other garments in the machine. Pouring in a color safe bleach solution can introduce some oxidizing molecules and stop this from happening.

Is it okay to use percarbonate on its own at 105°F (or 40°C) to get rid of germs?

No—with no activator, this is too cold for percarbonate to have any effect on germs and bacteria during an ordinary-length wash cycle.

Can I use percarbonate without an activator at 120°F (or 50°C)?

You can, but it won't be as effective. The rule of thumb in chemistry is that the speed of a chemical reaction cuts in half with every 18°F (10°C) drop in temperature. Broadly speaking, that means the percarbonate will take twice as long to do its job, and preprogrammed wash cycle times don't allow for this kind of tinkering.

I bought a percarbonate detergent that is supposedly effective at 85°F (or 30°C). Can this be true?

It most likely contains an activator. The same answer to the question above applies here: The chemical reaction is slower at this temperature, so the product won't perform as well as it would at 105°F (or 40°C).

I use a liquid detergent for colored clothes. Do I need to add delicate bleach as well?

I would recommend using delicate bleach if you're washing at 105°F (or 40°C) or lower, as the bleach removes stains, kills germs, and makes sure your clothes won't smell bad if you intend to dry them indoors.

Which products contain percarbonate?

Percarbonate is often added to powder detergents that claim to be powerful stain removers. It should be listed as an ingredient on the bottle (although it's sometimes hard to find). Liquid detergents never include percarbonate, but many of the powder additives sold as bleaching or antibacterial agents intended for use alongside liquid detergents do contain this substance. Most manufacturers have specific product lines that include percarbonate.

Can I buy percarbonate on its own?

Yes—some large grocery stores carry plastic containers of percarbonate, which often come with their own measuring cups. I buy mine this way, as I don't like the added fragrances in some prepackaged formulations. The hardware stores I go to for citric acid also sell percarbonate, and it can be easily sourced online as well.

How do I know if the percarbonate I bought contains an activator?

For reasons unknown, some manufacturers produce percarbonate but don't add an activator. My advice is always to check the ingredients list for SOBS, NOBS, or TAED. You need to know this since if there's no activator, the product won't work below 105°F (or 40°C), which severely restricts its usefulness.

Do I really need to add delicate bleach or percarbonate to a load of laundry along with my liquid detergent to prevent bacterial growth? What if there's already oxygen in the detergent?

Average wash temperatures used to be around 140 to 195°F (or 60 to 90°C), which was perfect for killing bacteria and mold. These temperatures have since dropped to 85 to 105°F (or 30 to 40°C), which is too low to reduce bacterial contamination. Therefore, if you can't dry your clothes right away, they can very quickly become a breeding ground for germs, with their associated odors.

Using a dryer helps prevent the proliferation of potentially smelly microorganisms. If you don't have one, then an oxygen-based bleaching agent is an effective alternative—because of their oxidizing properties, these bleaching agents can help reduce bacterial load. You can choose a delicate bleach containing hydrogen peroxide or a powder one containing sodium percarbonate. But if your chosen detergent already has an oxygen component (this should be listed on the product label), then you won't normally need to add anything extra.

Are sodium percarbonate and baking soda the same thing?

No—they are two completely different substances, and baking soda has absolutely no whitening or antibacterial properties. I will explore baking soda in more detail in chapter 11.

8

Laundry

We're so used to being able to wash our clothes whenever we want—even every day or multiple times a day—that it can seem strange to think there was once a period when doing the laundry was a time-consuming, labor-intensive chore that people did far less often, sometimes as little as once a month. You first had to boil huge quantities of water to soak your sheets, clothes, and anything else that needed cleaning, then wash it all with lye, soda ash, or soap (if you could afford it). After that, you had to spend ages beating and rubbing the fabric to remove the stains, then rinse it and wring it out. Finally, everything had to be hung outside to dry, often on the lawn or spread over bushes.

It could take several days to get through all the dirty laundry. Women, who were typically assigned the chore, also had many other daily tasks to take care of, whether they were working in the fields, running the household, looking after the children, or preparing food. So they simply didn't have time to do the laundry too frequently, which meant that dirty linens, sheets, and towels would pile up in the run-up to laundry day. The wealthy, or at least those who were better off, would boast about how many sheets or clothes they owned, since it afforded them the luxury of hoarding dirty laundry, perhaps while waiting for a washerwoman to come and spend a few days at their house taking care of it all. Those who were less well-off simply didn't change their sheets or other linens very often. Only in more recent decades did people begin doing the laundry on a weekly basis. Laundry day was often Monday to allow plenty of time for clothes to dry before the end of the week (even if the weather turned out to be bad), perhaps so people could have a fresh outfit for church on Sunday or at least set aside their work clothes for a day and put on their Sunday best.

Now that we have the luxury of choosing when to take care of our laundry, let's talk about how best to do it.

DID YOU KNOW?

Separating clothes by color isn't a modern-day requirement that came along with the washing machine. People have always separated their laundry, especially when water was scarce and heating it took effort. In the past, they'd wash white items first: underwear, sheets, and any other garments that weren't usually all that dirty. Then, using the same water, they would wash colored clothes and fabrics that were sometimes much dirtier, like overalls and work clothes.

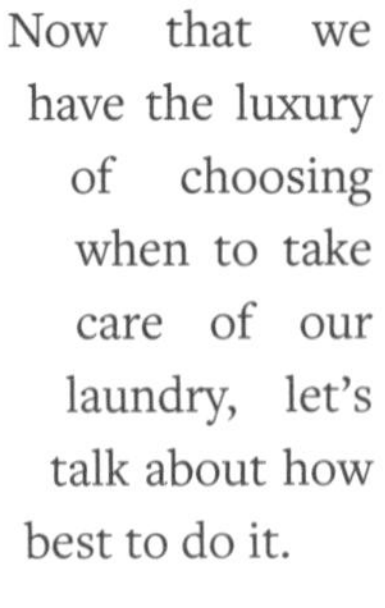

SEPARATING LAUNDRY BY TYPE OF WASH

It's important to separate garments that require different treatments—otherwise, we run the risk of damaging them, sometimes irreparably. The symbols on clothing labels help us out here by indicating the most vigorous treatment a garment can withstand. You'll sometimes also find these symbols on a washing machine's cycle settings.

Do not wash

Some garments can't be washed at home, perhaps because they're too delicate or covered with adornments that wouldn't survive a machine wash.

The standard symbol used to indicate recommendations for washing items at home is the washtub. Bleaching instructions are represented by a triangle, as we've already seen in chapter 6. The symbol for drying is the square, while ironing instructions are represented by an outline of an iron. Lastly, dry-cleaning instructions are identified by a circle.

If the washtub is crossed out, you can't wash the item at home either by hand or in the washing machine. This doesn't mean you can't treat individual stains locally, but you should always test the treatment on an area of the fabric that isn't visible first. To wash the item properly, though, you'll need get it professionally cleaned. The type of dry-cleaning required is indicated by an icon inside the circle symbol, but you don't need to worry about this: The dry cleaner will take care of choosing the most suitable solvent. However, if the circle is crossed out, this means the item can't even be dry-cleaned.

Wash less often

Now that we've ruled out items that shouldn't ever be washed, let's have a look at the ones we don't really need to wash. Yes, you read that right: Not all our clothes need to be washed every time we wear them. We've grown so used to the convenience of our washing machines that we don't think twice before chucking all our clothes into the laundry basket, even if a garment we've worn once or twice isn't visibly dirty. It can't hurt, we tell ourselves—but that depends. Each wash reduces a garment's life span a little more as detergents attack fabric, colors fade, and fibers degrade. So if possible, it's always best *not* to wash our clothes.

Things have changed drastically since our great-grandparents' time: We do less and less manual labor that causes us to sweat heavily and dirty our outfits, and the availability of hot water in everyone's homes has vastly improved personal hygiene, meaning there's less greasy residue on our clothes. Are you really sure that polo shirt or blouse you've worn for only a day needs to be washed? Maybe just leave it to air out for a couple of days. That's what I do whenever a sweatshirt I'm wearing smells like smoke because I've spent a bit too long near someone's cigarette. In this case, washing doesn't even help: Studies have shown that after 48 hours, the smell of smoke is reduced by the same amount whether you leave a garment outdoors or machine-wash it.

> **DID YOU KNOW?**
>
> Cotton and linen are the only textile fibers that are stronger when they're wet than when they're dry.

As a rule of thumb, here's what I do: The further a garment is from my skin's sebaceous and protein residues, and the less visibly dirty it is, the more likely I am to air it out before throwing it in the washing machine. Of course, this doesn't mean you can wear the same sweater for a month straight: Washing a garment, with fabric softener if needed, also helps take care of it by refreshing its fibers, removing any lint and fibrils (very fine strands made of protein and cellulose), returning it to its original shape, and sanitizing it. And I'm not suggesting you stop changing your underwear or socks—but you should at least consider leaving certain items of clothing out in the open rather than washing them. This is especially relevant at a time when energy bills are skyrocketing and people are becoming increasingly aware of their actions' environmental impact. Every time you avoid an unnecessary wash, the planet thanks you!

Hand-wash

The symbol showing a hand reaching into a washtub means your garment should be hand-washed (or you should select a special washing machine cycle that simulates hand-washing). The temperature should never exceed 105°F (or 40°C), but some garments need to be washed at lower temperatures, so always check the label. To hand-wash a garment, dissolve your detergent in water, soak the fabric, and leave it to float, swishing it around gently. Don't wring it out, rub it, or stretch it, since it needs to be hand-washed to avoid the mechanical shaking and rubbing of the washing machine. Rinse the item and press down on it to get rid of any excess water, but don't twist the fabric. For sweaters, laying a towel out on a table will help: Place the sweater on top and press down on the entire surface with a second towel. Then, leave it to dry, following the instructions on the label.

Machine-wash

If a garment can be machine-washed, there will be dots or numbers in the washtub symbol. These indicate the maximum temperature that the fabric can handle, with the numbers referring to degrees Celsius. Obviously, there's nothing stopping you from washing a garment at 105°F (or 40°C) even if the symbol says 140°F (or 60°C)—just don't do the opposite.

Water temperature isn't the only parameter you need to keep in mind: The mechanical energy required to clean garments depends on the friction between the clothes, how full the drum is, and the amount of water being used.

If a fabric needs special treatment to avoid wrinkles, there will be a horizontal line under the washtub symbol. This type of garment does best in the "permanent press" washing machine cycle. If you can't select that cycle, you should be careful not to overfill the washing machine, even if this means leaving it half empty. Also, either avoid the spin cycle completely or keep it short to reduce the risk of creasing and crumpling the fabric.

If there are two lines under the washtub, this means the garment should be washed on the "delicate" cycle. In this case, the washing machine's drum shouldn't be more than one-third full. Settings designed for delicate fabric generally use more water than the standard cotton cycle and work with back-and-forth rather than spinning motions to reduce the friction between pieces of fabric.

DID YOU KNOW?

If you use less detergent than the recommended amount, depending on the dirtiness of your clothes and the hardness of your water, you may end up with grayish deposits on the fabric. However, using too much detergent doesn't necessarily clean your clothes any better and also causes unnecessary pollution.

KEEPING COLORS BRIGHT

If you're keen to protect the colors of your clothes, you need to avoid several things: the fabric becoming degraded, the dyes getting physically removed from the fabric or destroyed, and other colored garments bleeding their hues onto everything else in the washing machine drum. Sadly, wearing clothes means that damage from things like repeated contact between your body and some parts of a garment more than others is inevitable—clothing fibers degrade bit by bit, which causes consequences like color fading. Fabric is subject to mechanical wear and tear even

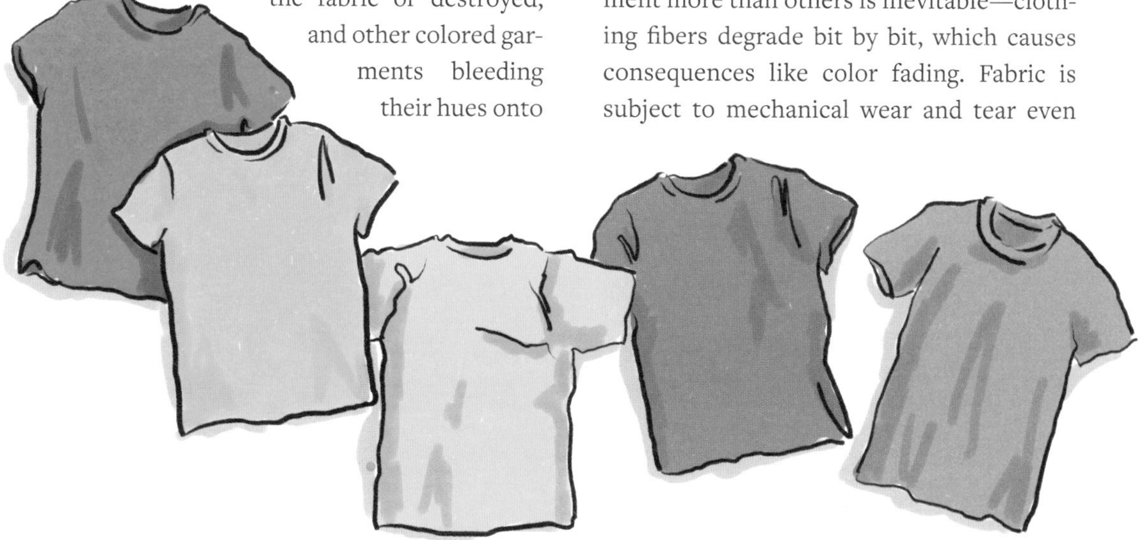

while it's being washed, especially if you don't select a delicate cycle, and this rubbing can also cause the colors to fade. This is why it's best to wash T-shirts (especially those with printed designs) and jeans inside out, with the colored side protected from mechanical abrasion.

Bleaching substances, especially chlorine-based products, can also cause colors to fade. This is especially true if you mistakenly use classic bleach instead of delicate bleach on colored fabric. If this does happen to you, unfortunately, there's nothing to be done: The colored molecules have been destroyed.

Laundry Checklist

- ☐ Keep dirty laundry in a dry, well-ventilated place while waiting to wash it. Don't leave it sitting there for too long.
- ☐ Always read the label and follow the instructions indicated by the symbols, which tell you the harshest treatment a garment can handle.
- ☐ Separate out garments that require different treatments (such as wool). Divide them up according to color and the maximum temperature they can be washed at. (Remember that you can always wash garments at lower temperatures than the maximum indicated on the label.)
- ☐ Separate out whites from colors and divide colored garments into two groups: light colors with warm, intense tones (like yellow, red, and purple) and dark colors with cool tones (like blue, brown, and black). It's best to wash new colored garments separately at first to prevent the color from running.
- ☐ Use a detergent suitable for the type of laundry (for example, it's best to use a special detergent for colored fabric).
- ☐ If you put clothes that are already wet in the washing machine (for instance, if you soaked them separately first), don't fill the drum more than halfway.
- ☐ Pretreat stains, but first check if colored fabrics can withstand the treatment by testing it on an area that isn't normally visible.
- ☐ Fill the washing machine drum according to the instructions on your machine. Some cycles require a nearly full load, while for others, the drum should only be filled halfway or less.

You may live in a place where the water supply is rich in chlorine composites, which are added to keep it drinkable and free of microorganisms. Although the amount of chlorine in the water is very small, after repeated washes, it can still fade the colors of garments that are made with particularly delicate dyes. Bleaches containing hydrogen peroxide that are designed for frequent use can help, as they react with the sodium hypochlorite in the water and cancel it out, producing oxygen.

Detergent is the main culprit when it comes to color fading, though. For synthetic fabrics, dye is usually added during the fiber production stage, but for cotton, it's applied to the surface, sometimes even after the weaving process. Detergents are designed to remove foreign materials from fabric surfaces, so it's hardly surprising that some of the dye comes off after multiple washes. The degree and speed of the discoloration depends on many factors, including the aggressiveness of the surfactant in the detergent and its pH level.

Detergents intended to protect colored fabrics and reduce this risk were launched onto the market in 1991. As well as having a lower pH and containing surfactants that are less aggressive toward fibers, they include neither optical brighteners nor traditional bleaching agents. They also solve another problem: If a bit of dye comes off the fabric, it can bleed onto another garment—like the infamous red sock that ruins an entire load. Detergents for colored garments contain an ingredient (often a polymer like polyvinylpyrrolidone or a molecule derived from it) designed to stop any dye from one fabric from landing on another. The polymers capture the rogue dye molecules on contact.

The same mechanism is applied in "color catcher" sheets, but in a much less efficient way. These are pieces of fabric or another material soaked in a substance that blocks dye if it lands on them. However, they have a weakness: Dye molecules are far less likely to land on the sheet than on a white T-shirt, which has a much larger surface area. A polymer dispersed throughout the laundry solution could prove more effective. Still, if that colored sock comes into direct contact with your favorite white T-shirt, there's nothing to be done.

Some products promise to make colors brighter, and they probably contain the enzyme cellulase (check the label). As a result of wear and tear, repeated washes, and the effects of detergents, microfibrils come off cotton fibers, and if there are lots of these microfibrils, they're visible to the naked eye—they look like tiny hairs or little balls. At the microscopic level, these microfibrils diffuse light and make the fabric's color look duller than it actually is. Limescale residue and dirt deposits can also lurk in these little hairs, making colors appear less vibrant. The enzyme cellulase detaches microfibrils, allowing color to be seen without any interference once again. Watch out, though: Cellulase doesn't work right away, so it takes several washes for it to restore color to its former glory.

There are some ingredients that make it more difficult for dyes to come off fabric in the first place. The downside is that they also make it harder to remove dirt from that fabric. In recent years, detergents designed specifically for black fabric have also been appearing in stores.

KEEPING WHITES FRESH

I bet you have some T-shirts or other garments that used to be bright white, but have now turned a sort of gray or yellowish color. I'm sure you'd love to restore them to their former glory, right? This isn't always possible, unfortunately—it depends on why the fading happened in the first place.

Over time, after multiple trips through the washing machine, white garments lose their original brightness and start to look washed-out or turn a different color. Chemical analysis suggests this may be due to metal ions slowly becoming lodged among the fabric's fibers. These metals can come from water or from products we use on our bodies, such as deodorants or cosmetics. Sebum from our skin also settles into fibers and acts like glue for dirt and proteins that are released from our bodies. Additionally, insoluble residues from soaps and detergents can get trapped between fibers and microfibrils after being precipitated by the calcium ions in water. These residues all make white (as well as other colors) look less bright.

A normal detergent partially removes these materials, but it can't get rid of them completely—to do that, you need a detergent containing some extra ingredients that we've already come across. Cellulase breaks down microfibrils, allowing detergent to reach fibers and remove any grease, while sequestrants remove color-altering metals and proteases get rid of proteins stuck to fibers. No detergent on the market can do all of this in one go, however—if the product delivers on its promises, it will take several washes to see the results.

WASHING SPECIAL MATERIALS

Wool and silk

These fabrics are made of proteins, and their molecules can be very delicate. They don't respond well to high temperatures, so never wash wool above 85°F (or 30°C). It's like what happens to egg whites when you heat them up: The proteins denature and their structure changes, and once that's happened, there's no turning back. (Yes, there is such a thing as baked wool, but it's been specially felted before getting this treatment.) Don't bother trying to fix heat damage to wool or silk with baking soda as some online "experts" recommend.

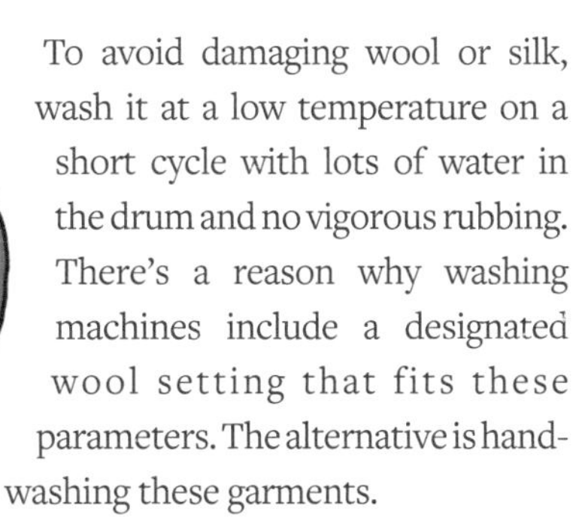

You don't need to only watch out for issues with temperature, since proteins are also highly sensitive to pH (both acidic and basic). Unfortunately, most detergents have a very basic pH, which is why, if you mistakenly put a wool sweater in the washing machine along with everything else, you can end up ruining it. It's even worse if the detergent contains proteases (enzymes specifically designed to destroy dirt proteins). When proteases come across wool or silk, they think it's a huge lump of dirt and attack it mercilessly.

To avoid damaging wool or silk, wash it at a low temperature on a short cycle with lots of water in the drum and no vigorous rubbing. There's a reason why washing machines include a designated wool setting that fits these parameters. The alternative is handwashing these garments.

Synthetics

When grease gets stuck to polyester fabrics, it's one of the most difficult types of dirt to get rid of. In contrast, it's relatively easy to remove grease from natural fabrics like cotton. This is because polyester fibers are hydrophobic, so they cling to grease more tenaciously and make it harder for water to get through. As a result, synthetic fabrics often smell bad even after they've been washed because the greasy, smelly substances are almost all still there.

One way to tackle this problem is to add polymers with both hydrophilic and hydrophobic regions to the detergent formulation.[1] These polymers stick to the synthetic fibers, making the fabric easier to wet and allowing the detergent to work. For this to be effective, however, the polymer must have already been deposited on the fibers, so it takes several washes to see the results.

1 One commonly added polymer is polyethylene terephthalate (PET).

GETTING RID OF STAINS

Detergents can get rid of most of the dirt on our clothes, but not all of it. Alas, many of the substances in food and drinks have intense colors and a nasty habit of staining our garments. Here are just some of the stains that detergents can't remove completely: coffee, tea, wine, chocolate, fruit, blood, lipstick, ink, rust, and grass.

Detergent's troubles exist because not all dirt is the same—so to help us clean as effectively as possible, it's worth taking a quick look at stains' composition. When we say "stains," we're referring to any colored substances that show up clearly on fabric or other surfaces, even in minute quantities, and can't be removed with ordinary detergent. Stains are often localized to a very small surface area and nearly always need to be treated locally, followed by a thorough wash if possible.

There are endless tips online about removing stains from laundry and other surfaces. I've found myself searching for tricks for removing yellow highlighter from a white wall, for instance, or getting chocolate ice cream off a couch with a nonremovable cover. When I originally planned this book, I intended to include some of these suggestions in this chapter. But I had a change of heart, because if you think about it, the number of possible stains is so high that, when you multiply it by the number of possible surfaces, there are too many results to cover here. Even if I stuck to the most common situations, explaining what to do for each stain/surface combination would take up far too many pages. Also, this wouldn't fit the scientific approach I take in the rest of the book, where I try to explain how to analyze a problem before offering specific practical advice. Analyzing the type of dirt you're trying to remove helps you determine whether the advice you find in magazines or online makes

STAIN-REMOVAL BEST PRACTICES

It's almost always a bad idea to heat stains, either during the pretreatment or drying phase. There's a risk this could permanently attach the stain to the fabric. If the stain's still there after you've pretreated and machine-washed a garment, try treating it again without drying it, perhaps using another approach.

Also, it's not a good idea to wait multiple days before tackling a stain. Don't rush unnecessarily, though. If you stain your shirt at a dinner party, you're unlikely to have any stain remover with you, and pointlessly scrubbing the mess with sparkling water or some other fanciful concoction can just make things worse by spreading the dirt even more.

Rust stains on fabric

Rust doesn't just form on metal objects—you can sometimes find it on clothing, too. Maybe it was a button that did it, or perhaps you got your shirt dirty while handling a rusty object. Rust dust is particularly tricky to remove because it's not greasy, which means a detergent won't touch it. Also, it isn't oxidizable (since it's already oxidized) and enzymes can't break it up.

To remove this type of stain, first suck up as much rust as possible from dry fabric using a vacuum cleaner with a nozzle—this is easier if the stain is recent. Next, try a 5-percent citric acid mixture, since citric acid helps capture iron ions and bring them into solution. Place a cloth under the stained fabric to absorb the liquid, then dab at the stain repeatedly. The idea is that the citric acid dissolves the oxide and captures the iron, which is carried away before being absorbed by the cloth underneath. After that, wash the garment in the washing machine.

Some people online suggest using a generic descaling product (the kind that works on taps) directly on the stain, but I wouldn't recommend this, as the dyes this product contains could stain the fabric even more. If nothing else works, go ahead and buy a special product, which generally contains specific chemicals designed to work on iron oxide—oxalic acid is the most effective ingredient.

sense (some of it does) or whether it's just nonsense (as is often the case). Sometimes, no DIY method works, in which case you have no choice but to buy a product specially designed for a specific stain or surface.

Remember that stains often contain two or more substances, which need to be treated one after the other using different approaches. For example, many inks are grease-based, so treating them with only a bleaching agent won't work, nor will just using a detergent. The same goes for tomato sauce stains. Now, let's think about how to classify different stains.

Water-soluble

These are the easiest stains to get out. They include sugar, toothpaste, shampoo, table salt, and other inorganic salts. (I frequently make crystals at home with the help of my chemistry set, so I sometimes find myself cleaning blue residues from copper sulfate and alum rock off my shirt.) All you need to remove this type of dirt is water, preferably hot. A regular machine wash or even a simple soak does the job.

Particulate

Stains formed by small solid particles can be very tricky to remove. They include coal dust, ash, residue from burned material (like burnt pizza bases or charcoal residue from a barbecue), soot, clay, other types of soil, and metal oxides. This type of dirt doesn't dissolve in water, isn't chemically attacked by detergent, and isn't oxidized by bleaching agents, so there's no point in pouring bleach or percarbonate onto it. Enzymes don't help, either. The only things that work are physically detaching the material from the fibers with mechanical energy and using a detergent or soap with a basic pH.

You should remove the dirt mechanically before washing the garment—with a bit of luck, its chemical composition won't make it particularly hard to detach from the fibers. When the fabric is nice and dry, try working on it with a powerful vacuum cleaner or brushing it to get rid of most of the dirt. Then machine-wash it, several times if needed.

If there's mud on your pants, resist the temptation to dampen it to get it off. Instead, wait until the mud has completely dried out, then brush it off—you may not even need to wash the pants at all.

WARNING! When you're handling organic solvents, be careful not to spill them. Always wear goggles and gloves, work outside if you can, and try to breathe in as little of the substance as possible.

Solvents can be very useful for tackling certain stains, but always check whether they're compatible with the specific fabric or other material you're working with. Acetone, for example, is perfect for removing glue, varnish, and ink residues, which is why it's used in nail polish remover, but it can easily damage anything made with acetate.

There's also a risk that the solvent will strip the material of its color. To be on the safe side, dip a Q-tip in the solvent and rub it onto a spot that's hidden from sight—for instance, inside a pants leg. Wait for the solvent to dry out completely before checking to see if there's any damage.

Some metal oxides can be attacked by acids, but check to make sure this won't damage the fabric's surface first.

Greasy

Greasy stains always contain a key oil component. It could be an edible fat like olive oil, butter, gravy, sauce, or yogurt, or it could be lipstick, mascara, or some other fat-based makeup product. It could also be sebum, lubricating oil, or wax. No matter what it is, when grease gets into textile fibers, it's very difficult to remove.

Obviously, water won't help—and no, don't bother with sparkling water (which I've seen recommended by some influencers). Bleach doesn't work because the stain's color doesn't come from substances that can be attacked by bleaching agents. For this type of stain, your first port of call should be surfactants, so you'll need a detergent. The basic pH of soaps and detergents helps remove grease, especially edible grease made up of triglycerides.[2] High temperatures also assist with removal.

You probably have various types of detergents at home, from laundry detergent to dish soap to shampoo. As you know, they don't all have the same properties, and some are more effective on oily stains than others. While it may seem odd to use

2 At a basic pH, triglycerides turn into diglycerides and monoglycerides, which are polar and thus more easily removed by the surfactants in detergents.

shampoo to treat a stain, the main ingredient is very often sodium lauryl sulfate, a very effective surfactant for treating grease that is, unsurprisingly, used to make hair less greasy.

After treating the stain with detergent, you'll obviously need to rinse it off. This can be tricky at times—for instance, if you've stained a sofa that doesn't have a removable cover. In this case, one option is using an organic stain-removing solvent. There are several types available, and finding the right one for your specific stain often isn't easy. Ethyl alcohol is an organic solvent often found around the house, but unfortunately, it's not very effective against grease (and vinegar and baking soda don't help, either).

When I was first getting into chemistry as a kid, I was always on the lookout for new substances I could use in my experiments. In my mom and grandma's cupboards, I often came across organic solvents designed for removing stains from fabric. One example was trichloroethylene (but now, it makes me shudder to think that this carcinogenic substance with a whole host of contraindications was commonly used in our homes). There was also acetone, which is still used nowadays for removing nail polish and other types of paint. With a bit of patience, you can find other solvents that are very good at getting rid of stains, such as turpentine and glycerol.

DID YOU KNOW?

There's no such thing as a universal stain remover, since different types of stains require different pH values, bleaching agents, and temperatures. For example, with percarbonate and a bleach activator, wine stains come off most effectively at a pH of 9, while tea stains are most easily removed at a pH of 11.

If possible, it's best to apply solvents—including water—to the opposite side of the fabric from the stain. For example, if you need to remove a mark on your shirt caused by a washable marker, turn the fabric inside out before applying water to carry the pigment away. To be extra careful, after reversing the garment, place a thick white cloth under the stain and apply the solvent so that as it gets into the fibers, it's also absorbed by the cloth.

Greasy stains are particularly difficult to get off synthetic fabrics, as synthetics are hydrophobic and lipophilic (that is, attracted to fats). In this case, pretreating the fabric with a specific product definitely helps: A nonionic surfactant is ideal for removing oily stains from synthetic fabrics.

If the stain is on a table or another piece of furniture, you should follow the same advice I've given for clothing: Make sure you're using the right solvent by testing it on an area that's not visible. While I was writing this book, I stained my varnished oak table green with one of the many gel pens I use to annotate articles. I forgot to put the lid back on, and while it was hidden underneath a pile

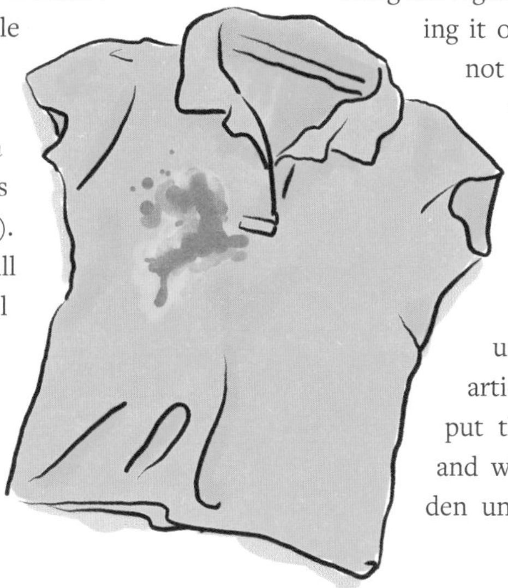

Stain removal with glycerol

Removing dried bits of food from an item of clothing can be tough, especially if the food is fat-based and thus not easy to get wet. In this case, glycerol—which you can find online or at a vape store—may help soften things up. Glycerol is an organic solvent that dissolves in water and comes off easily, and it can also be useful for removing traces of makeup or other fat-based cosmetics. Dilute glycerol with water by 50 percent, apply a bit of the liquid to the stain, and leave it to work for up to half a day. Once the stain has softened, treat it according to the type of dirt it comes from.

of paper, the pen left a few marks on the table. I only noticed a few days later, by which time the marks had already dried out. Soap and water did nothing. I tried using a Q-tip soaked in alcohol, but before tackling the green stain, I tested this on the edge of the table to make sure it wouldn't strip the paint off, too. The green stain came right off and the table was saved. You can also use acetone or turpentine to remove other types of ink. However, if a piece of furniture is precious to you, I wouldn't risk treatment for just a small stain.

Bleachable

The list of colored molecules found in stains is almost endless. They range from the green of chlorophyll (in grass and vegetables), to the orange and red of carotenoids (in tomatoes and peppers), to the pink, red, violet, and blue of anthocyanins (in blueberries, grapes, red cabbage, and strawberries), to the yellow of flavones and naphthoquinone (in henna) or curcumins (in turmeric, mustard, and curry), to the grayish-brown of humic acids (in tea and coffee).

As their name implies, bleachable stains can be removed with bleaching agents. The anionic surfactants commonly found in detergents can also partly wipe out most of these stains, but to get rid of them completely, you'll almost always need to use chlorine- or oxygen-based bleach. Oxygen-based bleaching agents are often included in powder detergents but are not usually found in liquid ones, which means you have to incorporate them yourself as additives. You can either put a bleaching additive directly into the washing machine drum or pretreat the stain by soaking it before washing.

The culprit responsible for colored stains is usually a molecule with a piece in its structure that chemists call a "chromophore" (which literally means "color carrier"). Bleaching substances oxidize and transform the chromophore, removing the color. The molecule that caused the stain may still remain attached to your shirt's fibers, but you'll no longer be able to see it. Bleaching agents can also work by turning colored molecules into more soluble ones, making it easier for a detergent to detach them from fibers.

Treatable with enzymes

This last vast and varied category includes all types of dirt that can be dealt with using specific enzymes.

For instance, all protein residues—such as grass, blood, body residue, meat juice, eggs, and milk—can be treated with proteases. This is most effective as a before-wash pretreatment at room temperature because heat (or just drying out over time) denatures proteins and can make them cling to fibers even more stubbornly. Locally pretreat the stain for ten to thirty minutes with a detergent containing protease, then machine-wash the fabric with detergent to get rid of any greasy or particulate residue. Bleaching agents can get rid of some of these stains, too—for example, bloodstains can be tackled with chlorine- or oxygen-based bleaches.

Enzymes known as amylases are effective against starch residues (like those from bananas, potatoes, cocoa, rice, and pasta). Often, the colored particles that make a stain visible are bound to the starch granules, which are in turn attached to textile fibers. Therefore, once you remove the starch, detergent easily gets rid of the rest. And as we've already seen, lipases can remove grease when working side by side with surfactants.

Many laundry and dishwashing detergents contain enzymes these days, but unfortunately, their labels don't always specify which ones. Although enzymes are sometimes included in liquid detergents, they appear in powder detergents more often and in greater quantities. This is because in powders, the enzymes are more easily protected from degradation by the other ingredients in the package, while preserving enzymes is much more difficult and costly in liquids.

Sunscreen stains

Some sunscreens contain molecules that act as filters against UV rays but can also stain clothes—in particular, avobenzone and oxybenzone. If you wash fabric with traces of a sunscreen containing these molecules, you can end up with yellow, orange, or pink stains. This is especially likely if the washing water is rich in iron and the detergent you're using has a highly basic pH. Chlorine-based bleaches only make matters worse, since they bind the stains to the fabric, and unfortunately oxygen-based bleaches don't help, either.

Some products' patents claim that you can get sunscreen stains out by locally applying an acid solution (such as citric acid) and pretreating with an amphoteric surfactant such as an amine N-oxide or cocamidopropyl betaine. So if you're desperate, look for a dishwashing detergent or shampoo that contains one of these surfactants (or at least some kind of betaine). Pretreat the stain for half a day, then machine-wash without any bleach, ideally using a detergent with a neutral or slightly basic pH.

THE WASHING PROCESS

Whether you opt for a trusty bar of soap or a synthetic detergent, and whether you're washing dishes or sheets, cleaning relies on four key factors: mechanical energy (from your hands or your washing machine), thermal energy, chemical action, and time, all facilitated by the fundamental power of water. The relative importance of these different factors depends on the type of cleaning required.

Each of the three actions—mechanical, chemical, and thermal—takes a certain amount of time to work, and they go hand in hand with one another: The hotter the water, the more effective the detergent, and the less time and mechanical energy required to wash an item. At lower temperatures, you either need more time or more detergent.

Mechanical energy

From a scientific perspective, washing means removing unwanted material from a fabric—usually with water, although dry cleaners also use organic solvents—through physical and chemical processes that rely on mechanical energy. When you hand-wash your clothes, the mechanical energy comes from your hands and arms (although once upon a time, some people also pressed clothes with their feet like they were making wine from

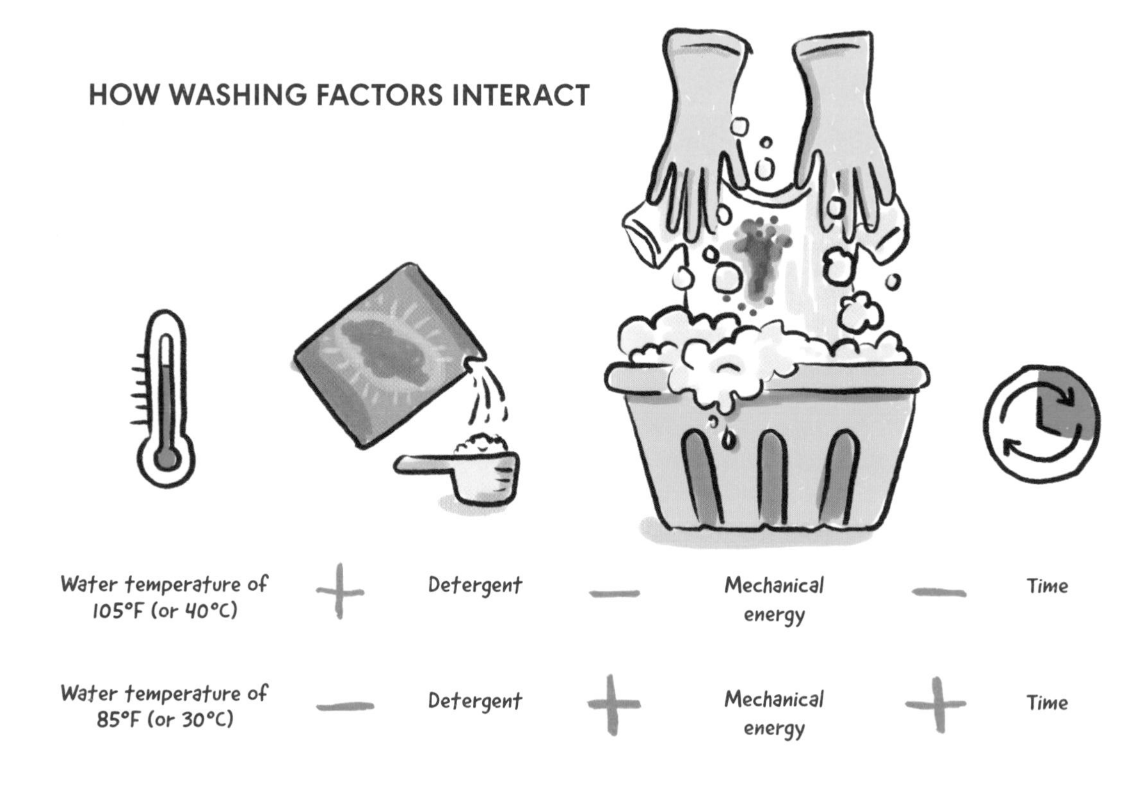

RECOMMENDED AMOUNTS OF DETERGENT FOR 10 LBS (OR 4.5 KG) OF LAUNDRY

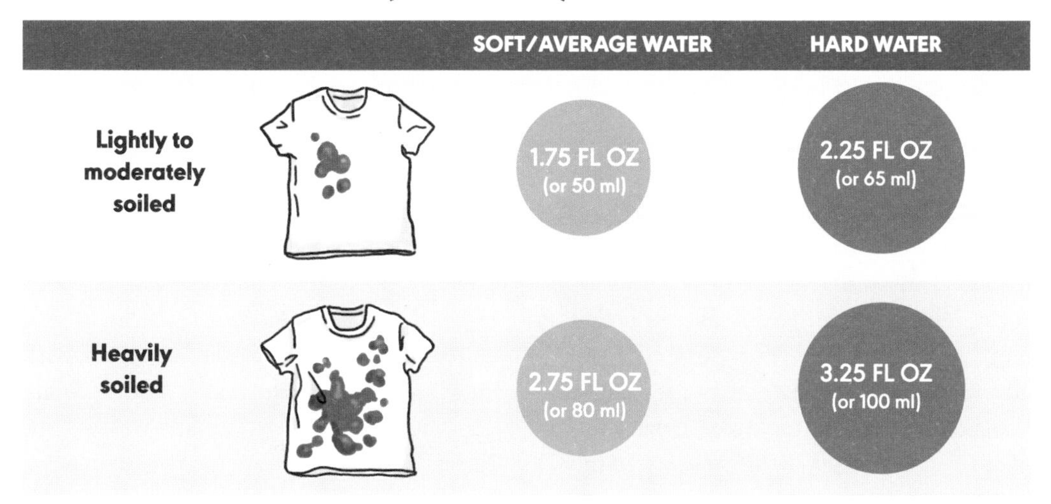

grapes). In a washing machine, it comes from the laundry rolling and rubbing against itself as the drum rotates.

More mechanical energy can be produced by hand-washing than by machine-washing, since with the former, you can choose to scrub laundry more vigorously if necessary. Also, if you hand-wash your clothes, you have the option of soaking them for an hour or more before washing, allowing water to seep deep into the fibers and thus making detergent more effective. With a machine-wash, you have to compensate for minimal (or even nonexistent) soaking time by using detergents that are far more complex than a lowly bar of soap.

Water temperature

In the past, people sometimes boiled laundry, and until fairly recently, most washed their clothes with the machine set to 195°F (or 90°C). Washing machines still include that setting, but it's hardly ever used nowadays. The 140°F (or 60°C) setting has taken its place, and these days, people regularly wash at even lower temperatures between 85 to 105°F (or 30 to 40°C) for energy-saving and environmental reasons.

As temperature decreases, so does the speed of chemical reactions: Roughly speaking, every time you reduce temperature by 18°F (10°C), you halve reaction speed. Because of this, washing machine manufacturers had to increase the duration of many cycles, and detergent producers had to significantly change their ingredients, which now need to do the same job at cooler temperatures that they used to perform at far hotter ones.

Water hardness

One thing people don't give much thought to when doing the laundry is how hard their water is. Be honest: How many times have you actually followed the instructions on detergent packaging? You may not have even noticed that there are at least two dosage instructions: one for very hard water, and another for soft or medium-hard water. We've already learned that when it's used in hard water, soap is very ineffective and leaves behind residue. To a lesser extent, this type of water also makes detergents containing anionic surfactants (that is, almost all of them) less effective. (Nonionic surfactants don't present this issue, but you won't generally find them on their own in common detergents.) When your water is hard, you have to compensate for the presence of high amounts of calcium and magnesium salts by increasing the amount of detergent you use.

Also, hard water leads to limescale gradually building up on the inner surface of pipes, reducing the water's rate of flow. As it accumulates, limescale traps dirt that passes through the pipes—like hair or grease—and forms a compact layer that's tricky to remove. It also builds up inside household appliances, which can be damaged in the long run if it isn't removed. While some modern dishwashers come with a softener designed to reduce the amount of limescale in the water, washing machines don't have anything of the sort. This means that limescale can accumulate not only inside the machine, but also on your clothes—destroying their softness, ruining their appearance, and eventually wrecking their fibers completely.

Limescale deposits can be removed, with a bit of effort if they're very dense (see chapter 3 for lots of ways to get rid of this substance). You can use a descaling product to do this for your washing machine. However, it's always better to prevent limescale before it forms, for at least three reasons:

1. Limescale is highly insulating, so it reduces the efficiency of water-heating elements if it's left to build up on them. Imagine covering your radiator with a layer of Styrofoam—you'd never be able to heat up the room in the winter. This analogy may seem a bit extreme, but it demonstrates the concept well: More limescale means less hot water.

2. Calcium and magnesium ions in hard water react with soaps and certain detergents, forming an insoluble substance that precipitates and no longer performs its cleaning job—and can accumulate on your clothes. This is why all washing machine detergent packages specify recommended dosages depending on the type of water you have at home.

3. Micro-incrustations of limescale settle into the fibers of your lovely, soft sweaters and scarves, making them lose their texture and turn stiff. After repeated deposits, their colors will also fade or turn gray.

DID YOU KNOW?

Detergents designed for industrial use or for businesses like hotels often have a different formulation from household products. Hotels, for instance, have to wash huge quantities of sheets and pillowcases that are generally not heavily soiled by cosmetics or sebum from our skin. Therefore, detergent manufacturers can adjust the mixture to make their product more effective, allowing their customers to use the smallest possible amount to tackle specific types of dirt.

PRODUCT SPOTLIGHT: **CALGON**

One of the most famous descaling products out there is Calgon, which has been on the market since 1933. The brand is now owned by the multinational Reckitt Benckiser. The name "Calgon" comes from the first letters of "calcium" and "gone," since it catches the calcium that forms limescale in washing machines and flushes it down the drain.

Formulations of this product vary by location and type (for example, depending on whether it comes in liquid, powder, or tablet form, or whether or not it also sanitizes). Let's take a look at the ingredients of Calgon's powder form.[3]

The first ingredient on the list (meaning it makes up the highest percentage of the overall product) is sodium sulfate, which has no descaling properties. Sodium sulfate is an inert ingredient that is widely used to bulk up powdered products, making it easier to measure dosages. Imagine if, instead of a measuring cup or tablespoon, the instructions told you to use a fifth of a teaspoon—you'd find it much harder to dose the product correctly. But without sodium sulfate, Calgon would still cost the same, since this ingredient is plentiful and very cheap.

In second place is the main anti-limescale ingredient, zeolite. Zeolite is a mineral that acts as an ion exchanger: Essentially, it captures calcium ions and (to a lesser extent) magnesium ions, and in return, it releases sodium ions. Zeolite's structure consists of silicon, oxygen, and aluminum atoms arranged in a three-dimensional framework with many cavities and empty channels on the inside. Because of the empty channels' size, some substances can get in but others can't, allowing zeolite to trap calcium. Both natural and synthetic zeolites are widely used as water softeners in powder detergents.

Then there's water, which is probably added to act as a solvent for one of the other ingredients. The next item on the list is polycarboxylate, which serves the same function as citric acid: capturing calcium ions with its acid groups. However, while citric acid has only three acid groups, in polycarboxylate, there are a potentially infinite number, all lined up in a row.[4] A molecule of citric acid (that is, citrate) can sequester a small calcium ion, while polycarboxylate can wind itself up like a thread and trap larger limescale particles, holding them in suspension. Through the same mechanism, polycarboxylate prevents limescale from forming on surfaces. Calgon also contains a surfactant, which helps keep the other components in solution in the washing water, and a clay called bentonite. Bentonite stabilizes emulsions and helps prevent the formation of limescale and dirt—it's also used in some fabric softeners. Lastly, you'll find all the usual perfumes and colorants.

---

3 You can check the ingredients of Reckitt Benckiser products at rbnainfo.com.

4 Because of the pH of the water in your washing machine, all these acid groups are negatively charged carboxylate ions.

Can I make it at home?

No. We don't have anything in our homes that performs the same functions as zeolite or polycarboxylate.

Having established this, for the real DIY enthusiasts among you, let's talk about the gel version of Calgon. It doesn't contain any zeolite, which would be difficult to suspend in a liquid since it's an insoluble powder. And instead of polycarboxylate, there's another polymer with the same job: sodium citrate (which, as we've seen, is very effective at sequestering calcium). As it's a liquid product, it also contains preservatives.

Sodium citrate is the salt produced by a reaction between citric acid and an alkaline substance containing sodium. Citric acid is readily available online and in shops that specialize in cleaning products, and you'll even find it in some hardware stores. In principle, you could react citric acid with any of the following alkaline substances: lye, sodium carbonate, or baking soda. However, you can't just pour citric acid straight into your washing machine's detergent compartment. While detergent's alkaline conditions do transform citric acid into precisely the citrate we're looking for, this process also neutralizes at least some of the detergent, stopping it from doing its job. Therefore, you need to convert citric acid into sodium citrate separately.

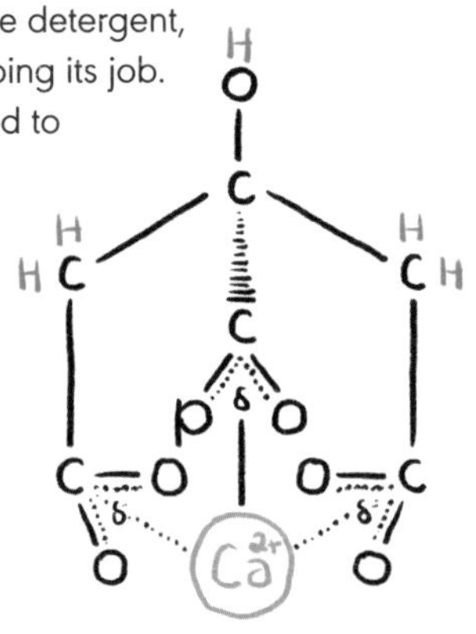

Of course, this isn't a chemistry lab manual, and I don't want to hand out instructions that could result in an accident if they're not understood completely and followed to the letter. I'm sure only a few of you are actually interested in making your own homemade descaler purely for personal enjoyment—but if you are, you can use the information in this section to write down a chemical equation for producing citrate and calculate the amount of baking soda needed to do so. For those of you who don't feel up to the task, there's no need to mix substances, since chemists have done the job for you already and you can buy the fruits of their labor at any grocery store.

A citrate ion bonding a calcium ion

WHAT'S IN CALGON?

INGREDIENTS	FUNCTION
Sodium sulfate	Inert, adds volume
Zeolite	Water-softening agent, traps calcium ions
Water	Solvent
Polycarboxylate	Sequestering and anti-depositing agent, inhibits limescale's formation
C12-13 Alketh-7	Surfactant, acts as an emulsifier
Bentonite	Stabilizer and softener

Water amounts

Washing machine cycles use not just varying temperatures, but also remarkably different amounts of water. The crucial parameter to keep in mind is the ratio between the weight of the dry laundry and the water in the drum. Some of this water is soaked up by the fabric, while the rest remains free—in a standard cotton cycle, only the bottom quarter of the drum is filled with water, which is almost entirely absorbed by the dry laundry. The more water there is, the more the garments float around and the more space there is between them, which reduces the friction they experience. For this reason, delicate or wool cycles use up to three times as much water as cotton cycles.

Another factor that affects the mechanical energy transmitted to the fabric is the movement of the drum, which is programmed to rotate and change directions periodically. A combination of limited water, long rotations, and frequent changes of direction interrupted by limited pauses generates strong mechanical action. In contrast, a large amount of water relative to the amount of fabric, limited short rotations with less frequent changes in direction, and longer pauses create less friction, which explains why this type of cycle is called "delicate."

DID YOU KNOW?

Cotton fibers are made up almost entirely of cellulose, which can comprise as much as 97 percent of the fabric. Linen, on the other hand, has a cellulose percentage of 64 to 75, while for hemp, it's 64 to 70. The remainder is mostly hemicellulose and lignin. Since all these fabrics are mostly cellulose, they handle high washing temperatures well—but make sure you stick to the instructions on the label.

DID YOU KNOW?

The latest washing machines can weigh the laundry in the drum by rotating it a few times. This feature, known as "auto-sensing," allows the machine to determine the water level and washing time required for a particular load.

Fabric types

In general, we can classify fabric fibers into three categories: cellulose-based (cotton, rayon, linen, and hemp), synthetic (polyester, nylon, acetate, acrylic, and polyamide), and animal protein–based (wool, silk, and alpaca).

CELLULOSE-BASED

Cotton is one of the most important fabrics. Its fibers are very absorbent and become negatively charged when wet. It also gets dirty easily but, thanks to this negative charge, the alkaline pH of detergent gets rid of most of the dirt. The enzyme cellulase is very effective in helping detergent clean the fibers thoroughly. Cotton fibers handle high washing temperatures well and can be treated with chlorine- and oxygen-based bleaching agents.

SYNTHETIC

Synthetic fabrics like polyester, nylon, and acrylic have much smoother fibers than cotton, so it's harder for dirt to get in. However, acrylic fabric is hydrophobic (that is, it repels water), so greasy stains cling to it much more stubbornly. Unlike cotton, synthetic fibers are much worse at handling high temperatures, which can cause them to soften and even lose their shape. They are also much less resistant to attack by bleaching agents.

ANIMAL PROTEIN-BASED

Like other animal-based textiles, wool can be ruined if it's washed the wrong way. Because of this, it's sometimes specially treated to allow it to be machine-washed, but the washing still needs to be done very carefully using the correct detergent. If you're worried you might make a mistake with your favorite sweater, err on the side of caution and handwash it instead.

DID YOU KNOW?

After you've used your washing machine, it's a good idea to leave the door open for at least a few hours to get rid of any residual moisture. This helps prevent the formation of mold or bacteria. (Make sure the cat doesn't get in, though!)

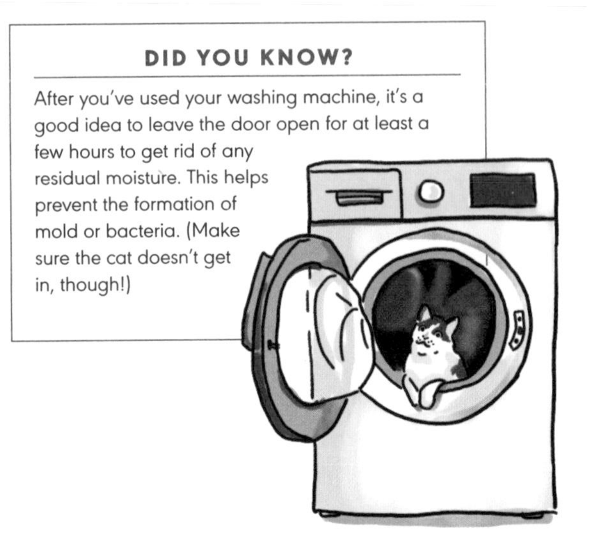

FABRIC SOFTENER

Why fabric gets rougher

Is there anything worse than getting dressed in the morning and finding that your T-shirt or sweater, which used to be so soft and lovely to wear, has become all stiff and scratchy? What's happened to it? You've always taken such good care of it, and you only just washed it—which is exactly the problem.

Every day, almost our entire body comes into contact with different fabrics for hours on end. Ideally, this contact needs to be pleasant rather than irritating. Natural textile fibers like cotton are rough, but when they're transformed into fabrics that are used to make our clothes, they go through various processes collectively known as "finishing." These treatments differ depending on a fabric's desired characteristics and can involve physical, mechanical, or chemical procedures. Before World War II, oil and grease were deposited onto fabric to make it less rough to the touch and increase the lubrication between fibers. Nowadays, specially designed substances, like silicones, are used.

The ways we treat fabric fibers can be compared to hair. Our hair is made from natural protein fibers called keratin that are normally coated with sebum, a fatty substance produced by our sebaceous glands. Sebum protects against the external environment, waterproofs, and lubricates. When we wash our hair with shampoo, we strip away that protective layer, leaving it exposed to the elements and stopping the strands from freely sliding over each other. Silicones are also the most widely used ingredients in shampoo to counteract this stripping. To restore that protection while waiting for sebum to build up again, we use conditioner. It works more or less the same way for textile fibers. Cotton is made up of cellulose rather than proteins, but after repeated washing and drying—especially if you use a clothesline or drying rack—the effects of the initial treatment wear off, leaving the fibers exposed and causing them to harden. Bit by bit, the cotton fibers unravel and form microfibrils. Once dry, these microfibrils remain upright, facing toward the outside of the fabric. And if your water is hard, limescale builds up, creating a sort of microscopic sandpaper effect that makes the fibers more strongly attracted to each other, causing them to lose their flexibility and slippery quality.

How fabric softener works

Fabric softeners aim to stop all this from happening. They're made with cationic surfactants, whose heads contain a positively charged ion that helps the molecule disperse in the water and bind itself to the fibers.[5] (As we've seen, detergents typically include surfactants that are anionic instead.) The surfactant also includes at least two long, fatty (and therefore hydrophobic) chains that lubricate the fibers. The positively charged surfactant is attracted to the fabric fibers, which usually become negatively charged in water (especially if they're cotton fibers). This provides a mechanism of action to get the softening process going, since the charged surfactant makes fibers repel each other, reducing the amount of friction in the overall washing process.

Fabric softeners work in both acidic and basic environments. An acidic pH is best, though, since it helps neutralize the residual alkalinity on the fabric caused by the detergent.

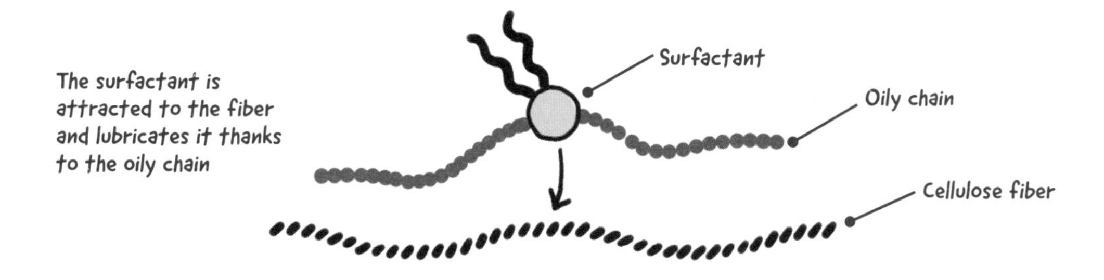

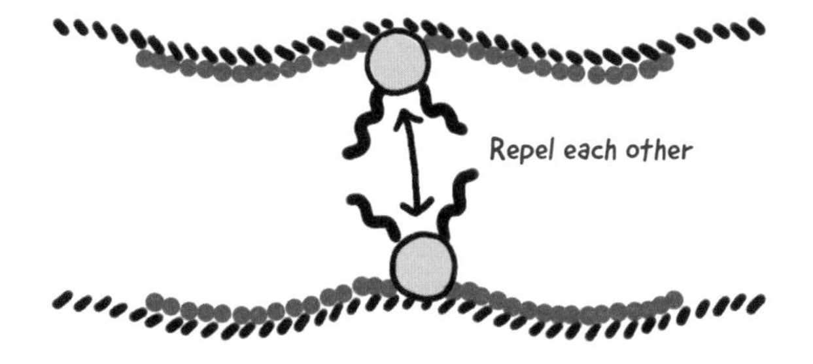

5 For many years, the only surfactant used was DHTDMAC (dihydrogenated tallow dimethyl ammonium chloride), known colloqiually as "quat," which was patented in the late 1940s.

Fabric softener concentration

Although fabric softeners all have more or less the same formulation, you'll find two different versions of these products on the shelves: conventional and concentrated. Don't be tricked by the name, though: Concentrated fabric softeners, which were first introduced in Germany in 1979, aren't simply a concentrated version of conventional ones. Instead, they have a slightly different formulation based on a higher concentration of active ingredients—between 4 and 8 percent in conventional products and 12 and 30 percent in concentrated products.

Concentrated softeners offer some advantages for both companies and consumers. They take up less space, are less expensive to transport, and result in less plastic that must be recycled. However, they're also more difficult to dose, which may cause you to end up using more than needed, causing waste as well as consumer dissatisfaction when the bottle runs out sooner than expected. Some consumers are also wary of concentrated softeners because, at least subconsciously, they associate a product's effectiveness with the amount used. And of course, concentrated products may also cost more per unit of weight (but not per washing unit) than conventional ones, which can make people wary of purchasing them.

The history of fabric softener

The fabric-washing revolution began in the 1950s, when the first synthetic detergents emerged along with the first household washing machines. These innovations cause more damage to clothing fibers by stripping off their finish more easily—and hanging clothes out to dry rather than tumble-drying them makes things even worse. Although dryers are now common in the United States, they're far less prevalent in Europe, making fabric texture issues much more of an problem there. At the same time, the fabric of choice in Europe is still cotton, while synthetic fibers are popular in the US. Synthetic fibers are less likely to lose their softness but tend to have built-up static electricity when they come out of the dryer, causing garments to stick to one another and sometimes even give you a small electric shock on very dry days. The first fabric softeners—launched in the US market in 1957 and in Europe in 1963—solved both the stiffness and static electricity problems. Since the 1970s, softeners have contained at least two active ingredients: the main cationic surfactant and a performance-enhancing substance known as a co-softener.

Fabric softener marketing

"A sensory experience," "stimulates emotions," "floral freshness," "a world of fragrance": These are just some of the descriptions I've seen on a well-known fabric softener's website, and I think they provide a fair idea of how these products are marketed. These phrases could just as easily refer to a cosmetic. The focus is always on softness, fragrance, and tactile sensations—that is, on emotional and sensory properties rather than functional or technical ones. It's no coincidence that, if you take a look at the ingredients of the bestselling fabric softeners, you'll find that they contain more than twenty ingredients added for the sole purpose of creating a specific scent: bergamot and nettle-leaved bellflower, purple orchid[6] and blueberries, peony and white tea, ylang-ylang oil,[7] lotus dreams, forest dew, cotton clouds, pink amethyst[8] . . . the list goes on and on. The ad copy's accuracy is up for debate, though—what exactly does "yearning for the Maldives" or "breath of the Dolomites" smell like, after all?

There's probably no other product out there for which each company has so many different versions, distinguished only by their scent. Indeed, tests have shown that the main reason many consumers opt for a specific fabric softener is its fragrance and how long the scent lasts on their clothes. It's no coincidence that companies invest considerably in testing out many different fragrances to create a distinctive perfume, even though they can get away with far less work on normal detergents. Creating the perfect perfume is no easy task, since it needs to permeate throughout the liquid and last even after the fabric dries completely.

ENVIRONMENTALLY FRIENDLY FABRIC SOFTENERS

In 1990, the Danish and German authorities decided to put DHTDMAC—the cationic surfactant used in the vast majority of the world's fabric softeners up to that point—under observation. They were concerned by the surfactant's very poor biodegradability and asked manufacturers to find alternatives. By 1993, the use of DHTDMAC plummeted by 70 percent, and a similar but more biodegradable molecule was used.[9] Since then, dozens more environmentally friendly surfactants that can be used in fabric softeners have been developed.

---

6 Goodness knows how the scent of a purple orchid differs from that of a red or orange one. I imagine it isn't the same as the scent of a black orchid, however, which is used in a competing product in combination with patchouli oil.

7 Which, honestly, I was completely unaware of (as many of you probably were, too). However, Wikipedia informs me that it is an essential oil made from *Cananga odorata*, a plant belonging to the *Annonaceae* family that's one of the main exports of the Comoro Islands.

8 I'd like to give a prize to anyone who can describe what a pink amethyst is supposed to smell like, given that it's a mineral.

9 This molecule is called an esterquat. For curious chemists: Essentially, two ester groups are introduced into the two saturated hydrocarbon chains to make the molecule more biodegradable.

What do fabric softeners really do?

Despite what advertising tries to convey, fabric softeners are needed not just for their emotional benefits, but also for the maintenance of garments and their fibers. Have you ever tried roughly estimating the amount of money you've got stored in your closet? Give it a go one day to get an idea of the capital you have dangling on hangers or folded up in your drawers. If a softener could extend the lives of your clothes, allowing you to wear them for an extra year or two, you'd be pleased, right?[10] Just think about how much money you'd save. That's not the only aspect worth considering, though, so let's have a look at some other benefits we haven't discussed yet.

DID YOU KNOW?

Consumers often (wrongly) associate high viscosity—essentially, how difficult it is for a liquid to flow freely—with how concentrated its active molecules are. People are wary of very fluid products because they associate thickness with higher concentration, so companies add substances specially designed to increase viscosity. But remember, using unnecessary ingredients anywhere along the product supply chain always leads to some level of pollution and wasted resources.

DRIES FASTER

The surfactants used in softeners are hydrophobic, which means that during the rinse cycle, they ensure that fabric fibers absorb less water and release the water deposited on them more easily. If you use a dryer, this leads to faster drying times and energy savings. It's difficult to say exactly how much, but some studies suggest there is a 14-percent reduction in drying time, with a similar decrease in energy consumption.[11]

WARNING! Never pour liquid fabric softener directly onto an item of clothing—doing so can leave a stain. Also, never put detergent and fabric softener in the washing machine at the same time, since they will react with one another, forming substances that can also stain your clothes. The two products are incompatible, which is why you should always add softener to the designated tray or wait until the rinse cycle to pour it in. By that point, most of the dirt and detergent have already drained away.

MAKES IRONING EASIER

Cotton garments often need to be ironed while hot to get rid of any wrinkles in the fabric. A fabric softener makes ironing easier and more efficient: Because surfactant molecules act as a lubricant between fibers, the temperature of the iron helps them to stretch out easily as they slide across one another. This reduces ironing time by as much as 10 to 20 percent, which may not sound like a lot—but just think about how much energy this saves.

PROTECTS FIBERS

Fabric softeners' lubricating function also results in less friction, which means less wear and tear. It's

10 Actually, a garment's life shouldn't be measured in years, but rather how many times it's been washed and the number of days it's been worn.

11 Kuo-Yann Lai, ed., *Liquid Detergents*, 2nd ed. (London: Taylor & Francis, 2006.)

important to remember, though, that this protection only applies while the fabric is dry, not while it's being washed. To protect your delicate clothes from the washing machine's mechanical action, it's important to select a corresponding delicate cycle.

Negative effects of fabric softeners

Of course, there's always a downside to everything, so let's have a look at the potential disadvantages of using a fabric softener. Adding too much, which is always a risk with concentrated products, can leave fabric feeling greasy due to too many lubricating molecules settling on the fibers. In the long run, the hydrophobic molecules that fabric softeners deposit can also reduce cotton's wettability. This can be problematic, particularly for towels, since it means the fabric absorbs water more slowly. Also, fabric softeners can interfere with the optical brighteners contained in some detergents, making white clothes appear less vibrant. Finally, some people may have an allergic reaction to fabric softeners' fragrances or other components, resulting in redness or itchy skin.

PRODUCT SPOTLIGHT: **LIQUID FABRIC SOFTENER**

After water, the main ingredient in liquid fabric softener is the softening molecule itself, which acts as a lubricant. While fats and soap derivatives were sometimes used before the invention of synthetic surfactants, cationic surfactants are most popular nowadays. Each company has its favorite, but they all have virtually the same structure: two or more long fatty chains linked to positively charged quaternary nitrogen ions. These chains are what give each surfactant its unique characteristics, and their lengths and types depend on the raw material used to synthesize the surfactant—which could be palm oil, mutton or beef tallow, or some other natural source. (If you want to avoid animal products in your fabric softener, check the ingredients to make sure it doesn't contain dihydrogenated tallow dimethyl ammonium chloride, aka DHTDMAC.)

Additionally, fabric softeners contain an alcohol (usually isopropyl alcohol), which dissolves the surfactant in water during the production phase. As we've learned, they also include a long list of fragrances used to create their perfume. Some products contain other molecules that make the fabric feel soft to the touch, such as silicones. There's also always a viscosity regulator to ensure that the liquid is thick enough for consumers to feel satisfied with the product. And last but not least, as with all liquid cleaning supplies, there's a preservative.

Enzymes, specifically cellulases, are the latest ingredients to be added to some softeners. They break down the final glucose molecules that form cellulose chains. This is important because, as I've already mentioned, fabric forms microfibrils over time: little balls of fluff that look unsightly on your sweaters and T-shirts. Dirt and limescale build up in these balls, which results in nasty micro-incrustations in the long term. You can remove these balls with a lint roller, allowing future washes to get rid of the incrustations and leave the fabric feeling softer.

Can I make it at home?

No. There aren't any household ingredients that can lubricate and maintain fabric fibers.

But this isn't the only function of fabric softeners, even if it is the main one. When doing laundry, our great-grandmothers sometimes added a bit of vinegar to the rinse water, especially if they used soap to wash their clothes. The acetic acid helped neutralize any alkaline residue left by the soap and the negative charges absorbed by the fibers, so in this sense, vinegar had a limited capacity to restore fabric's softness. Modern fabric softeners usually have an acidic pH between 2.5 and 5, allowing them to do the same thing.

Since acids are only partially effective and act more as a pH neutralizer than an actual fabric softener, if you do want to use one in your laundry, I recommend citric acid instead of vinegar, which is overrated. Although it's rare, citric acid is actually found in a few commercial fabric softener formulations. Unlike vinegar, it doesn't smell and doesn't damage your washing machine's seals, and it's excellent at trapping the residual calcium ions released by limescale removers during the wash cycle. You can add a 15-percent citric acid solution to the fabric softener tray or during the rinse cycle. Sodium citrate is even better at trapping calcium, but it requires an alkaline pH, while a softener must be acidic.

DRYING

Now, it's time to dry your laundry. You have two options: Firstly, as people have been doing since mankind first began washing clothes, you can hang it somewhere. The second, more modern choice is using a dryer to dry garments with hot air while they're kept in constant motion due to the continuous rotation of the drum. There are a few advantages to this method compared to using an indoor drying rack. High temperatures and shorter drying times stop bacteria and fungi, which can cause unpleasant odors, from growing on fabric. Also, continuous movement prevents magnesium and calcium carbonate residues from forming large crystals on fibers, causing fabric to become rough and stiff. Of course, all this comes at a cost in terms of energy usage.

Clothing labels include standardized symbols indicating how a garment should be dried. The typical shape for drying instructions is the square, and if there's a circle inside it, that means you can use a dryer.

As in the other cases we've seen, this symbol crossed out means the garment can't be dried in a tumble dryer.

The number of dots indicates which heat setting you should use: Three dots mean high, two mean medium, and one means low.

There are symbols for drying without a machine, too, but they're much less common. They're made up of one or more lines inside the drying square. A horizontal line means the garment should be laid flat, while a vertical line means it should be hung on a clothesline or hanger. Three lines (horizontal or vertical) mean the garment should be drip-dried to prevent creases from forming. To drip-dry a piece of clothing, take it out of the washing machine while it's still dripping wet.

A diagonal line added to the symbols I've already described means the garment should be dried in the shade or indoors rather than in the sun. For example, the rightmost symbol below means that the item should be dried flat without being exposed to the sun.

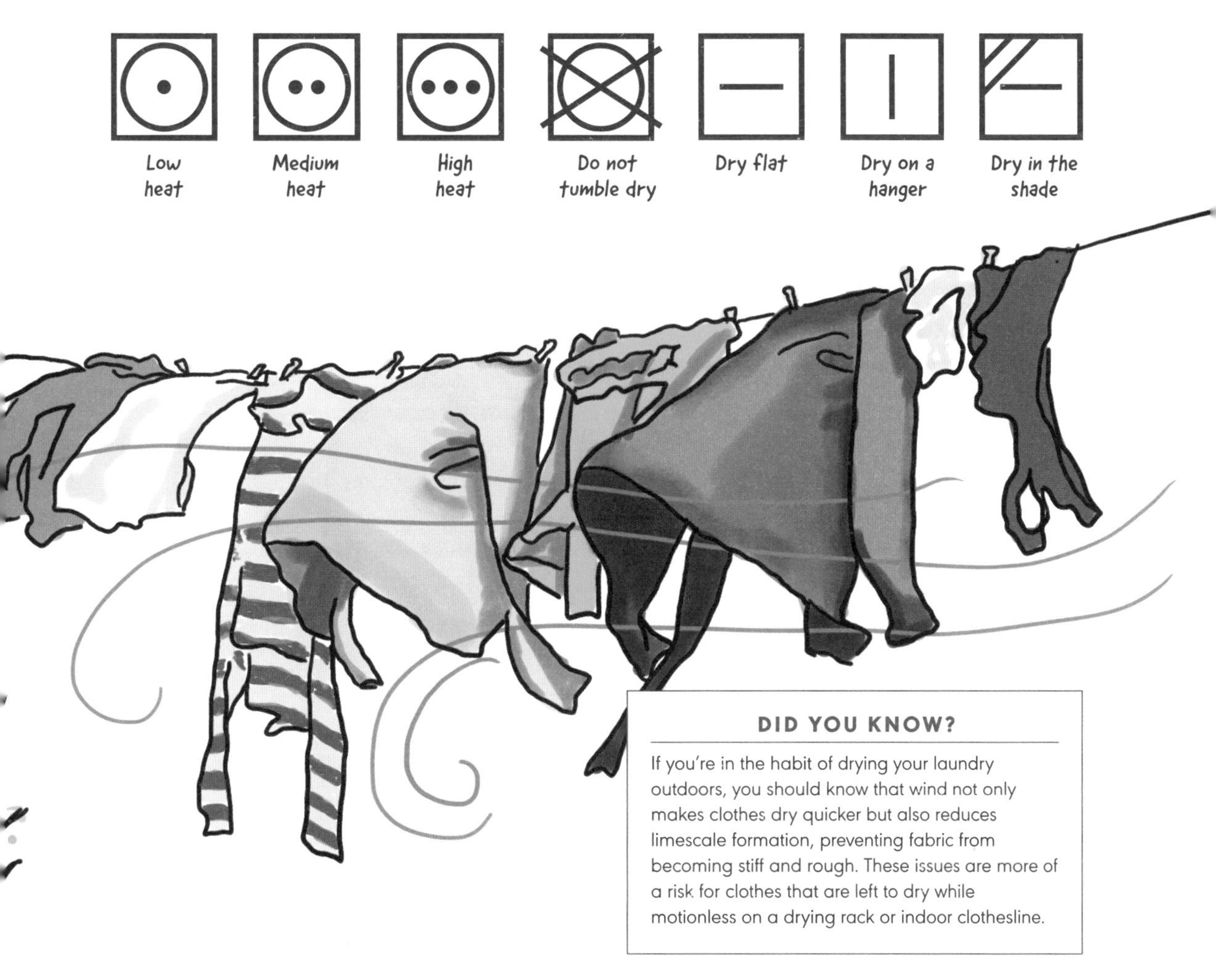

DID YOU KNOW?

If you're in the habit of drying your laundry outdoors, you should know that wind not only makes clothes dry quicker but also reduces limescale formation, preventing fabric from becoming stiff and rough. These issues are more of a risk for clothes that are left to dry while motionless on a drying rack or indoor clothesline.

IRONING

Lastly, let's take a look at the symbols for ironing, which is represented by the outline of an iron with dots inside.

As always, if the iron is crossed out, this means the garment cannot be ironed.

Three dots mean the garment can be ironed at a maximum temperature of 390°F (or 200°C). Generally speaking, only cotton, linen, and denim can be ironed at such a high temperature, and even then the garment should be dampened when necessary and possibly even turned inside out for this process.

Two dots indicate a maximum ironing temperature of 300°F (or 150°C).

A single dot indicates a maximum ironing temperature of 230°F (or 110 °C). This usually applies to synthetic fabrics.

9

Dishes

Every day, in billions of homes worldwide, there are plates, bowls, pots, pans, silverware, baking trays, and many other dishes to wash. There's also a range of products for tackling this chore, divided into two main categories: those for use in the dishwasher and those for cleaning dishes by hand. The two aren't interchangeable—each has different functions and a specific composition for its washing type—so don't try to use dish soap in the dishwasher or scrub your plates with dishwasher detergent!

Many factors affect washing efficiency, so you shouldn't be surprised if the methods and products that work for your friends aren't as helpful for you. The most important of these factors are washing method, mechanical action, type of dirt, type of dish being cleaned, detergent composition, and water quality.

As we wash our dishes, it's important for us to be mindful of the resources we're consuming. In its 2021 report on the global state of water, UNESCO predicted that by 2050, there would be a 55-percent increase in the overall demand for water.[1] At the same time, studies have shown that washing dishes by hand consumes more than three times the amount of water as using a dishwasher.[2] Ever since the earliest dishwashers appeared on the market, manufacturers have been perfecting their products, making them easier to use and increasing their environmental sustainability (often motivated by energy- and resource-saving regulations). Besides water, our homes are responsible for about one fifth of the world's energy consumption.[3] So overall, reducing domestic use of water and energy by running our households more efficiently is a worthy goal.

Can we wash our dishes more efficiently without compromising on hygiene and convenience? Science's answer is yes, we can—if we're willing to change our habits a bit. Many studies have shown that switching from washing the dishes by hand to using a dishwasher results in lower energy and water consumption, CO_2 emissions, and costs.[4]

Although on average, using the dishwasher is more sustainable than manual washing, if you do this incorrectly, it can reduce the environmental benefits. The three most common errors, which I'll discuss in more detail later, are rinsing dishes before putting them in the dishwasher, running the dishwasher before it's full, and using settings that wash your dishes at an unnecessarily high temperature.

---

1 UNESCO, *The United Nations World Water Development Report 2021: Valuing Water* (Paris: 2021).

2 Rainer Stamminger, Angelika Schmitz, and Ina Hook. "Why consumers in Europe do not use energy efficient automatic dishwashers to clean their dishes?" *Energy Efficiency* 12, no. 3 (March 2018): 567–83.

3 U.S. Energy Information Administration, *Annual Energy Outlook 2020* (Washington, DC: U.S. Department of Energy, January 2020).

4 Lotta Theresa Florianne Schencking and Rainer Stamminger. "What science knows about our daily dishwashing routine." *Tenside Surfactants Detergents* 59, no. 3 (March 2022): 205–20.

TYPES OF FOOD RESIDUE

The dirt we need to remove from plates and pans is not the same as the dirt found on our clothes or in other parts of our homes. Unless you eat nothing but plain boiled rice (in which case, I'd advise you to see a dietitian), after pretty much every meal, you deal with residue from three large groups of macronutrients: fats, carbohydrates, and proteins.

Fats are probably the most common grime we wash off our dishes. When you finish your pasta and there's still a bit of sauce left on the plate, these leftovers might consist of liquid vegetable oils or of animal fats that solidify at room temperature after cooking. Vegetable fats can also be solids (like cocoa butter or coconut oil), and animal fats can be liquids (just think about what's left in the pan after you cook salmon). But in the end, all this fatty residue is easily removed by the dishwasher as long as the water temperature is high enough. If the temperature is too low, the fatty residue is more difficult to get rid of because dishwasher detergents don't contain the most suitable surfactants for cooler water (unlike the liquid we use to wash dishes by hand). If those better surfactants were included in dishwasher detergent, the appliance's powerful churning action would create too much foam, which can damage its components.

Carbohydrate residue commonly comes from pasta, rice, beans, and vegetables, while protein residue comes from eggs, dairy products, meat, fish, and some other vegetables, like spinach or avocados. These types of dirt are usually difficult to remove. However, the chemists who formulate detergents have started using special enzymes that work alongside detergent's high pH and the dishwasher's mechanical action to get rid of carbohydrates and proteins.

Finally, aside from these three main residues, dishes can also acquire stains from drinks like coffee, tea, and wine. Fruits and vegetables produce colored stains, too—some culprits are carotenoids from tomatoes and anthocyanins from berries.

FOOD RESIDUES	STAINS
FATS (sauces, vegetable oils, animal fats)	**DRINKS** (coffee, tea, wine)
CARBOHYDRATES (pasta, rice, beans, vegetables)	**FRUITS AND VEGETABLES** (tomatoes, berries)
PROTEINS (eggs, dairy products, meat, fish)	

HOW DISHWASHERS WORK

Dishwashers clean dishes by combining three factors: mechanical energy from spraying water, thermal energy from hot water, and chemical action from detergent that's transported by water. Surprisingly, this last force isn't the most important one: Mechanical action accounts for about 85 percent of washing ability, followed by water temperature and then by the surfactants in detergent. Washing time varies depending on how different appliance settings combine these factors.

A dishwasher's mechanical action comes from its rotating arms, which spray pressurized water from their nozzles onto the dirty dishes to physically remove food residue. Obviously, the more arms the appliance has, the more efficient it is, but efficiency also depends on the quantity and pressure of water used. If you're like me, you've made the mistake of forgetting to put detergent in the dishwasher and noticed that at least some of the dishes were still fairly clean when the cycle finished—that's mechanical action in practice.

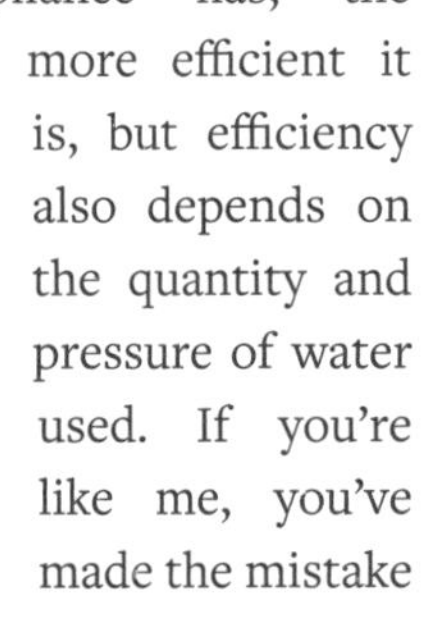

Almost all types of dirt are most easily removed with water that's hotter than room temperature. High temperatures in the dishwasher also help to dissolve detergent and, most importantly, melt fats that solidify at room temperature. However, excessively high temperatures can ruin delicate china and crystal, so don't put your best dinnerware set through a washing cycle unless you've confirmed that it's dishwasher safe.

Detergent's job is not just helping to remove dirt, but also controlling any inorganic ions (like calcium and magnesium) in the water. As we've learned in chapter 3, these ions become a problem if your area has hard water and you don't use a water softener. Dishwasher detergents contain surfactants that are specifically selected to reduce surface tension and produce little to no foam (so your foaming dish soap has no place in the dishwasher). These products also include additional ingredients like enzymes, bleaching agents, odor-removing components, and rinse aids.

DISHWASHERS VERSUS HAND-WASHING

When surveyed, consumers give various reasons for not owning or using a dishwasher. Often, people bring up two incorrect beliefs: They think hand-washing cleans dishes more thoroughly or uses fewer resources. Some live in a rented space that didn't come with a dishwasher installed, while others are concerned with practical issues like initial cost or lack of room in smaller kitchens. Other people say that since they use only a few dishes at a time, a dishwasher is less efficient than hand-washing. Emotional reasons like fear of potential water leakage and disinterest in buying yet another electrical appliance are also sometimes cited. Overall, up to one third of the people interviewed in one study were happy with their current dishwashing situation and saw no reason to change it.[5]

Here, I'll explain why the two myths I mentioned above turn out to be incorrect when you consider the scientific evidence.

Myth: Hand-washing cleans dishes more thoroughly

People often have strong opinions about this subject based on what they're used to doing at home, but let's make things clear right away: Using a dishwasher guarantees a higher level of cleanliness and hygiene than washing dishes by hand.

The first scientific study comparing mechanical and manual washing from a hygienic point of view was carried out in 1947, when two scientists inspected about one thousand New York restaurants, gathering thousands of samples of washed dishes.[6] At the time, only 17 percent of restaurants owned a dishwasher. The scientists' microbiological tests showed questionable hygiene across all restaurants, but the hand-washing results were always clearly inferior: For example, only 10 percent of hand-washed glasses passed the microbial hygiene test compared to 36 percent of those cleaned in a dishwasher. The study concluded that using a dishwasher was definitely better than washing by hand, but there was still room to improve the appliance's efficiency.

You might be thinking, "Yeah, sure—but do I really want to compare my home kitchen

5 Rainer Stamminger, Angelika Schmitz, and Ina Hook. "Why consumers in Europe do not use energy efficient automatic dishwashers to clean their dishes?" *Energy Efficiency* 12, no. 3 (March 2018): 567-83.

6 H. J. Kleinfeld and L. Buchbinder. "Dishwashing Practice and Effectiveness (Swab-Rinse Test) in a Large City as Revealed by a Survey of 1,000 Restaurants." *American Journal of Public Health and the Nation's Health* 37, no. 4 (April 1947): 379-89.

to a restaurant?" Hygiene regulations for restaurants have definitely greatly improved since 1947, but in 1956, a second study that focused on household kitchens confirmed the results: Plates, pans, and silverware washed by hand have greater bacterial contamination than those washed in a dishwasher.[7] A more recent study from 1983 also showed a higher level of bacterial contamination in hand-washed dishes—the source was often the dish towels used for drying.[8] On the other hand, mechanical washing successfully removed almost all bacteria present, even when the plates had been artificially contaminated with E. coli cultures.

Myth: Hand-washing uses fewer resources

Although modern dishwashers directly derive from the first models designed in the late nineteenth century, since then, their cleaning capacity and energy efficiency have changed greatly.[9] Average energy consumption has gone from 1.43 kilowatt-hours (kWh) per dishwasher cycle in 1999 to 1.035 kWh in 2005, and today, the "Eco mode" cycle on my dishwasher consumes less than 0.8 kWh despite lasting for four hours. And while dishwashers from 1950 consumed over 600 watts of electricity, today's appliances use just a few dozen—still mostly for heating water. (I've just bought a new dishwasher, and while my old one had an exposed heating element that controlled the final drying phase, the new one's heating element is covered and the dishes are dried by hot water and steam instead in a more eco-friendly way.) Also, dishwashers' average water consumption has been cut in half, reaching about 3.5 gallons (or 13 liters) in 2005,[10] while the most efficient modern dishwashers use less than about 2.5 gallons (or 10 liters) per cycle.[11]

We can still do better, though, in ways that don't depend on dishwasher innovations but rather on our own behaviors. I say this because other studies show that even people who use the dishwasher every day often don't do it correctly. In particular, many people rinse dishes under the faucet before putting them in the dishwasher—which is completely pointless and uses more water than necessary (more on this soon). What's more, consumers tend to overuse shorter, higher-temperature dishwasher cycles that consume more energy instead of letting the appliance choose the time and temperature with the energy-saving "Eco mode" available on some dishwashers. Even with these inefficient mistakes, some studies show that water consumption per item washed is half that of washing by hand, so using a dishwasher is still superior when it comes to protecting the environment.

7 Elaine Knowles Weaver, Clarice E. Bloom, and Ilajean Feldmiller, *A study of hand versus mechanical dishwashing methods*, Ohio Agricultural Experiment Station (Wooster, OH: The Ohio State University, May 1956).

8 M. A. Blackmore et al. "A Comparison of the Efficiency of Manual and Automatic Dishwashing for the Removal of Bacteria from Domestic Crockery." *Journal of Consumer Studies & Home Economics 7*, no. 1 (March 1983): 25-9.

9 Francesco Rosa et al. "Dishwasher history and its role in modern design." *2012 Third IEEE HISTory of ELectrotechnology CONference (HISTELCON)* (Pavia, Italy: IEEE, September 2012): 1-6.

10 Rainer Stamminger. "Daten und Fakten zum Geschirrspülen per Hand und in der Maschine." *Seifen, Öle, Fette, Wachse* 3 (2006): 5.

11 Christian Paul Richter. "Automatic dishwashers: efficient machines or less efficient consumer habits?" *International Journal of Consumer Studies* 34, no. 2 (March 2010): 228-34.

DISHWASHER TIPS

Don't rinse dishes before putting them in the dishwasher

Prewashing or rinsing dishes before loading them into the dishwasher is not only unnecessary—it's a waste of resources. All dishwasher cycles include an initial rinse, and dirt is removed during the washing phase anyway. Let your dishwasher handle the job it was designed for! All you need to do is throw away any solid food scraps beforehand.

Many consumers still believe that dishwashers can't clean dishes properly if they haven't been rinsed first, and some explain this by saying that their plates and other items don't get clean without this extra step.[12] In truth, the quality of wash strongly depends on how carefully you maintain your dishwasher, which includes choosing cycles and detergents according to the manufacturer's recommendations, cleaning the filter, and removing any limescale blocking the spray holes in the arms. If your dishwasher isn't in perfect working order, the solution is not to prewash dishes by hand, but to clean it so it can perform its function more efficiently (and possibly to switch detergents, which we'll discuss later).

Load the dishwasher correctly

You know that booklet with tons of pages of dense writing that you've never taken out of the plastic cover? Well, now it's time to read through those manufacturer's instructions. I bet you'll find out you're making numerous mistakes while loading your dishwasher, like facing plates the wrong way or standing glasses upright in the rack instead of tilting them. If you look inside an empty dishwasher and imagine the trajectory of the water that sprays out of the nozzles on the rotating arms, you'll understand how to position things to make sure all your dishes get clean. In general, pans should be placed facedown in the lower rack so that the water jets hit the dirty surface, while plates should be loaded vertically with the dirty side inward.

Choose the right cycle

Modern dishwashers offer a wide range of cycles, yet many people almost always use the same one, perhaps occasionally picking a more intense cycle when their dishes are particularly dirty. This isn't a good idea—you should adjust the cycle based on the specific load you're washing.

Nowadays, many dishwashers have a button that allows the appliance itself to select the washing time and temperature using water sensors that can tell how dirty the dishes are. And if you want to save energy and water, your dishwasher may have an "Eco mode" that washes at a lower temperature. Sure, it may take longer, but do you really need clean dishes in just one hour?

---

12 Christian Paul Richter. "Automatic dishwashers: efficient machines or less efficient consumer habits?" *International Journal of Consumer Studies* 34, no. 2 (February 2010): 228–34.

Typically, dishwashers have four phases: prewash (usually with just water), main wash (with detergent), rinse, and dry. Prewashing occurs only in some cycles—typically those for very dirty dishes or for pots and pans—and deals with food residue that's easier to remove. In the main wash phase, the high temperature and mechanical force of the water jets coming from the arms do most of the work, helped by the detergent. In the rinse stage, the dishwasher removes residue, helped by the rinse aid—which then prevents etching or clouding, especially on glasses, in the drying phase that follows.

DISHWASHER DETERGENT

Ingredients

Although we remove the same kinds of dirt whether we clean dishes in the dishwasher or by hand, the types and properties of detergents we use for each task differ greatly. Detergent has many ingredients because it has to act on many varieties of dirt while performing several other functions. Let's take a look at what dishwasher detergent contains and why, in my opinion, you shouldn't make it yourself.

SOFTENING AGENTS

These substances are also known as "builders." The presence of excessive calcium and magnesium ions has a negative effect on detergents' efficiency and causes limescale to accumulate on dishes and inside the dishwasher itself. Therefore, various softening agent formulas exploit the properties of certain molecules (like sodium citrate or some polymers) to capture limescale and prevent it from depositing on your pots and pans.

Vinegar is not a softening agent, so save it for your salad dressing rather than pouring this liquid into the dishwasher before running it. Sodium carbonate is technically a softening agent and a source of alkalinity, but it's not commonly included in dishwasher detergents because it reacts with calcium ions, making the insoluble calcium carbonate fall onto the dishes. Sometimes sodium citrate is used, but it's more expensive than other substances.

Softening agents can also help dispel dirt once it's removed from the dishes. When your water is hard, grease can get deposited on glasses and form an opaque film, but these agents help prevent this. Silicates are also widely used for this purpose, but large quantities of some silicates can have the unintended effect of corroding glass and making it opaque instead.

ENZYMES

As we've learned, enzymes are special proteins constructed to destroy other molecules or break them down into smaller molecules that can be eliminated more easily. The first use of enzymes in detergents dates to the mid-1960s. They're not always specified in ingredient lists, but if you see a word ending in "-ase," that's almost certainly an enzyme. Proteases, for example, destroy protein residue on dishes, while amylases do the same to starch residue. Enzymes are delicate ingredients, and a skilled product formulator must pay attention to their stability, which is easily impacted by a detergent's other ingredients.

BLEACHING AGENTS

Just like in the washing machine, bleaching agents are used in the dishwasher to get rid of colored stains. These chlorine- or oxygen-based substances chemically destroy the color-transmitting parts of molecules without actually removing the molecules themselves. Bleaching agents based on sodium hypochlorite are present in various liquid or gel detergents, but they are incompatible with enzymes. Oxygen-based bleaching agents like sodium percarbonate are more often used in dishwashing powders and tablets.

DISPERSANTS

These are polymers (like polyacrylics or polymethacrylates) or sodium silicates that prevent film formation and crystallization on glasses, stopping them from becoming opaque.

SURFACTANTS

There's a big difference between the role surfactants play in dishwasher detergent and their role in dish soap. The latter contains a large amount of surfactants, which need to clean dirt efficiently with repeated hand-washing action. In a dishwasher, removing dirt is done with not only mechanical action, but also water temperature and alkaline content (which saponifies fats, turning them into soap). So in this case, detergent's purpose is primarily increasing water's wetting ability and preventing dirt from getting redeposited on dishes.

Nonionic surfactants are often used in dishwasher detergent because they tend not to produce a lot of foam, and in fact, they typically inhibit the formation of foam caused by protein and starch residues. In contrast, anionic surfactants create more foam, which decreases the pressure of water coming from the dishwasher's rotating arms, reducing its cleaning ability. Anti-foaming substances can be added to detergent formulations.

Dishwashing powders and tablets also contain alkaline substances like sodium carbonate (not sodium bicarbonate, so don't add baking soda to your dishwasher).

PRODUCT SPOTLIGHT: **DISHWASHER DETERGENT**

When we use the dishwasher, we expect our dishes to come out sanitized with no food residue, clouding, stains, damage, or bad smells. Unfortunately, this isn't always the case—which is why researchers have continued to improve products over time. Currently, dishwasher detergents' weak point is their relatively poor performance when they're up against cooked, dried, or burnt food residue on pots and pans. Chemists still have some work to do, so using elbow grease before loading these dishes into the dishwasher is the best alternative for now.

In other chapters, I've called out specific products and listed their exact ingredients. But when it comes to dishwasher detergents, it's difficult to choose one representative brand because there are so many differences in detergent formulas. So instead, I've made a non-exhaustive list of the types of substances you'll probably find in these products.

INGREDIENT	FUNCTION	EXAMPLES
Surfactants (nonionic)	Reduce the surface tension of water, emulsify dirt, produce minimal foam, and resist water hardness	Ethoxylated alcohols
Softening/ sequestering agents	Combat water hardness and maintain high alkaline levels in water to help remove dirt	Sodium citrate, sodium carbonate, sodium silicate, and polycarboxylates
Corrosion suppressants	Protect glasses, plates, and metallic machine parts from corrosion	Sodium silicate and zinc or bismuth salts
Bleaching agents	Remove stains and hygienize	Sodium hypochlorite and sodium percarbonate
Bases	Emulsify and remove fats	Sodium carbonate and sodium hydroxide
Enzymes	Destroy fats, starches, and proteins	Amylase, protease, and lipase
Special additives	Protect china and porcelain from patina deposits	Sodium aluminate, aluminum phosphate, and boron oxide
Thickening agents	Increase the viscosity of gel products for easier dosing	Polymers (like polyacrylates) and clays (like bentonite and laponite)
Perfumes	Cover the smells of food and other detergent ingredients	

Types

Now that we know what makes up dishwasher detergent in general, let's look at the types that are available for consumers nowadays.

POWDER

This oldest type of detergent is now the least common. Users complain that the powder often gets scattered around while they're filling the dispenser—not to mention that if you keep your detergent in the cupboard under the sink, humidity can make the powder clump together.

TABLETS

The earliest dishwasher tablets appeared in 1997 and were basically pressed powder detergents containing a substance to help them dissolve in water. Their formula is still much like that of the classic powder detergent, but they don't scatter powder or absorb humidity. Tablets are separated into individual doses, which may unfortunately be larger than your dishwasher requires. Also, they can have troubling dissolving during the wash phase, so you may find residue in the dispenser at the end of the cycle.

LIQUIDS AND GELS

The first liquid dishwasher detergent was launched by Colgate-Palmolive in 1986, then immediately imitated by other large producers of detergent like Procter & Gamble, Unilever, and Henkel. These early products were also essentially powder detergent dissolved in water with the addition of ingredients that increased viscosity. You had to shake the bottle to mix the ingredients before use—a bit like the bottles of milk that were delivered door-to-door in the twentieth century before the invention of homogenization. The first gel products appeared in 1991 to solve this problem.

Today's detergents marketed as liquids, gels, or liquid gels are all suspensions of solids dissolved in liquids with varying levels of viscosity. The advantage of liquid and gel (and powder) detergents is that their dosages can be adjusted based on the amount and types of dirt present. Unfortunately, they can't contain the same wide range of ingredients that can be added to powder detergent, and they are also less stable.

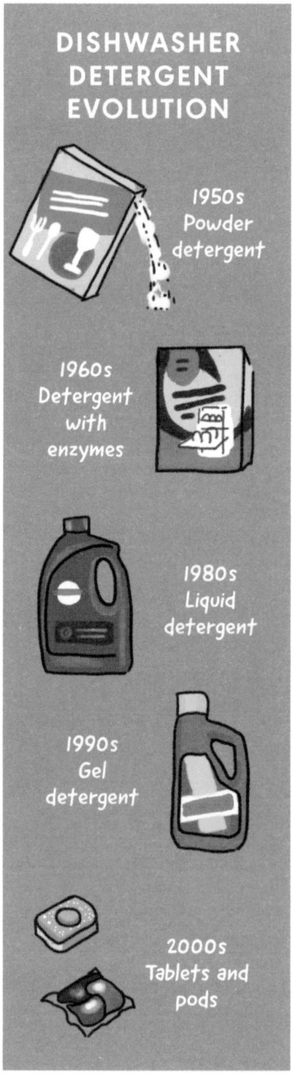

PODS

In 2002, we got the first liquid single-dose detergents stored in a film that dissolves to release the gel inside. Sometimes, these pods are divided into two or three compartments to separate incompatible substances until they are used. They compare to

DISHWASHER DETERGENT EVOLUTION

1950s Powder detergent

1960s Detergent with enzymes

1980s Liquid detergent

1990s Gel detergent

2000s Tablets and pods

liquid and gel detergents just like tablets compare to powdered detergent—they're predosed versions that are easier to use.

Be careful: Detergents have differences beyond their forms. Some contain enzymes, while others don't. And some are multifunctional (they're labeled something like "2-in-1" or "3-in-1"), while other traditional varieties have just one purpose. Multifunctional products contain not only detergents, but also other substances that help reduce calcium formation and prevent stains during the drying phase. However, a home water softener and ordinary rinse aid can do the same thing. If your water supply is hard and you don't use a water softener, you can try out multifunctional products to reduce the white marks that might appear on pans and glasses.

In April 2022, the Italian magazine *Altroconsumo* did a comparison of twenty-six dishwasher detergents: sixteen tablets, seven gels, two pods, and one powder.[13] They were assessed according to washing efficiency, shine, drying ability, ingredients, environmental impact, and price. When you look at the overall ranking, tablets significantly outperform gels, occupying seven of the first nine places, with a pod in third place and a gel in ninth. Of course, all rankings like this are subjective, and they can change over time as product formulations evolve. But even if we think about washing efficiency alone, it's not surprising that powder products (tablets are just compressed powder) perform better than gels. As we learned with laundry detergent (see page 86), gels come with restrictions because this form can make some ingredients difficult or impossible to include.

So, we can confidently buy tablets, right? Unfortunately not, because the three lowest-ranking products in terms of washing quality are two tablets and an eco-friendly powder (the most expensive of all twenty-six based on dosage).

Does that mean it's better not to throw away money on green products? No again, because at least two of the top five products have an excellent rating in terms of environmental impact. But keep in mind that just putting "eco" or "green" in a detergent's name doesn't give it a good environmental profile. Check the label for certifications like Green Seal, ECOLOGO, or EPA Safer Choice, and if you don't recognize a certification, look up its standards.

Should our choice of detergent be based on price instead? No, definitely not. Some of the higher-ranking products are inexpensive, and some of the pricier ones come in at the bottom of the pack.

What kind of dishwasher detergent should we buy, then? Well, I often try out various detergents to find what works best. At the moment I'm happy with gel products because my dishes aren't usually very dirty. But I'm definitely going to experiment with tablets, although I don't like the idea of not being in control of the dosage. If you want dishwashing perfection, you'll probably need to test a few different products, too.

---

13 "Detergents: how much do they impact on the environment? Discover this with the European CLEAN project," *Altroconsumo*, last modified March 8, 2021, altroconsumo.it/clean.

PRODUCT SPOTLIGHT: **JET-DRY**

Let's look at the composition of a simple rinse aid: Jet-Dry, which is made by Finish, a company owned by Reckitt Benckiser. Its first ingredient is water, and its second is a nonionic surfactant that lowers water's surface tension in the rinse phase. This causes water to collect on dishes as a thin film rather than drops, meaning it evaporates without leaving behind stains or clouding. Next is another surfactant, this time used as a hydrotrope, which prevents the ingredients from separating in the packaging. Then there's citric acid, which captures calcium and magnesium ions, and potassium sorbate, a preservative that's also used in the food industry, plus other preservatives called isothiazolinones. Zinc acetate prevents glasses from becoming corroded or opaque, and finally, there are the ever-present added colorants and perfumes.

Can I make it at home?

No, or rather, not entirely. The main ingredient for reducing water surface tension should be a specific nonionic surfactant, and we can't mix that in ourselves—at most, we can add an easily available metal ion sequestrant like citric acid. Some products contain up to 25 percent citric acid by weight, but they still also include surfactants and isopropyl alcohol. If you like, you can create a 10 to 15 percent solution by dissolving 10 to 15 grams (about half an ounce) of powdered citric acid in 100 milliliters (about half a cup) of deionized water, then pour it into the rinse aid dispenser.

A common piece of household advice suggests using vinegar as a DIY rinse aid. Since rinse aid is added during the rinse phase, the idea is that an acid introduced during this stage would react with the carbonates and bicarbonates present in the water to form carbon dioxide and leave behind soluble salts. In realty, the citric acid in commercial products is used for its sequestering properties, and the acetic acid found in vinegar is much less effective at capturing calcium. What's more, vinegar can corrode the rinse aid dispenser's rubber seals.

INGREDIENT	FUNCTION
Water	Solvent
Fatty alcohol alkoxylate	Nonionic surfactant
Sodium cumenesulfonate	Hydrotrope surfactant that prevents ingredients from separating
Citric acid	Hardness sequestering agent
Potassium sorbate and isothiazolinones	Preservatives
Zinc acetate	Corrosion and clouding preventative

TOMATO RED

One of the most troublesome residues to remove, especially from plastic containers that have been through the dishwasher, comes from tomatoes. The red of lycopene and other carotenoids changes from a charming component of many of our meals to a form of torture when we open the dishwasher and see all the plastic surfaces tinted with this color. (You've probably noticed that the same thing happens with curry.) Tomatoes' red pigments are not soluble in water—in fact, they are hydrophobic and try to stay away from it. But they are very easily absorbed by plastic, and high washing temperatures help these colored molecules to penetrate the surface. When they're embedded in plastic, they become even more difficult to remove with a surfactant or hide with a bleaching agent. Once, I used a pan to make tomato sauce and carelessly put it in the dishwasher without wiping it down—then had the terrible idea of washing all of my plastic food containers at the same time. I had to wash them again with a bleaching agent and a more aggressive detergent to remove the color.

CLEANING THE DISHWASHER

It might seem odd that you have to clean an appliance used for cleaning, but if this comes as a surprise to you, then you obviously haven't looked at the instruction manual. Wash after wash, some parts of the dishwasher fill up with dirt that negatively (and sometimes drastically) affects its performance. If your plates are coming out dirty or there's an unpleasant odor when you open the door, it's time to read up on dishwasher care.

You'll find out that the filter, which collects all the large residue that doesn't drain away, should regularly be removed, taken apart (if it's not just one piece), cleaned, and put back. This is a very simple procedure, but the exact details vary with different appliances. Also, if your water produces a lot of limescale, some of the holes on the rotating arms may become blocked. These arms can also be dismantled easily, and you can unblock the holes with a toothpick.

There may be crusted-on limescale in other areas of the appliance that you can't reach, though, and you can buy dishwasher cleaners to get rid of that. Dishwasher cleaners are used in an empty cycle (some appliances have specific cleaning cycles, but otherwise you can choose the hottest option). They remove both limescale residue and any other dirt, like bacteria and fungi, that settle out of sight in the dishwasher's depths and cause unpleasant odors.

PRODUCT SPOTLIGHT: **DISHWASHER CLEANER**

All the dishwasher cleaners whose ingredients I've checked contain the same components: one or more acids for descaling, one or more surfactants for removing stubborn dirt, colorant, fragrance, preservative, and water. The choice of surfactant varies from brand to brand, but the acid is invariably citric acid, and some products contain formic acid as well. The quantities of acids and surfactants vary greatly from one brand to another: For example, citric acid ranges from a minimum of 5 percent to a maximum of 25 percent. Products with more acid have fewer surfactants, and vice versa. Some cleaners also contain a sequestering agent and metal ions. Since fat residues commonly accumulate in the dishwasher, some varieties also have an alcohol used as a solvent.

Can I make it at home?

Almost. We're not interested in colorants, fragrances, and preservatives—the essential ingredients are surfactants and acids. If you have citric acid at home, use it to fill the detergent dispenser and start an empty wash cycle at the maximum temperature. This should clean out the limescale. We're just missing the active surfactant, which is especially important at a low temperature, which can increase the amount of residue left behind in your dishwasher. Splurge on good-quality detergent tablets and do another maximum-temperature wash with that.

Anti-clouding detergents

Some detergents claim to protect glasses from permanent clouding. One of the ingredients that may be added to these products is zinc salt. Calcium in the water tends to react with aluminum ions, producing calcium aluminate that's deposited on glasses and other dishes and is very difficult to remove. Using zinc salts reduces this phenomenon because although zinc aluminate also deposits on dishes, it's almost invisible, unlike the iridescent film of silicates. The zinc aluminate film protects glasses, preventing corrosion.

WHY GLASSWARE TURNS OPAQUE

You may have noticed that over time, some glasses that you clean in the dishwasher become progressively opaque, either in particular areas or all over. Tragically, once this has happened, it's impossible to restore glassware's original transparency and shine. This phenomenon has been reported and studied since the 1970s, and there are various causes, all linked to detergent.

If your glasses show a band of clouding near the upper rim, this might be a manufacturing flaw that alters the composition of the glass in those areas.[14] During machine-washing, alkaline detergent removes material from that part of the glass, leaving tiny holes that alter the diffusion of light, making the surface look clouded. If the glass is discolored all over, it might have undergone environmental corrosion for a few weeks after manufacture but before being packaged and shipped. One study has shown that if recently manufactured glasses with a certain composition are exposed to a damp environment during the day followed by condensation at night, this creates conditions that later lead to corrosion in the dishwasher.[15]

Temperature, mechanical action, and alkaline detergents (especially powder or tablets) may also work together to gradually dissolve some of the components of the glass. Another phenomenon is the deposition of a thin film of inorganic material (typically carbonates, phosphates, and silicates present in detergents) on the surface of the glass. You can actually prevent this from happening again or getting worse: Some multifunctional detergents or rinse aids offer protection against inorganic films getting deposited[16] or glass dissolving.[17] (I know reading a dishwasher product's label isn't the most exciting thing in the world, but I'm willing to pay the price of people looking at me like I'm slightly crazy when I spend several minutes taking in all the small print when I'm at the grocery store.)

In the end, though, if you have glasses that you're fond of, you should wash them by hand to prevent damage.

---

14 Klaus-Peter Martinek et al. "Local clouding of glass after machine dishwashing." *Glass Science and Technology* 78, no. 1 (2005): 12–7.

15 Melek Orhon, İlkay Sökmen, and Gülçin Albayrak. "Dishwasher corrosion of glasses." *Advanced Materials Research* 39 (April 2008): 317–22.

16 Using sulfonates, polymethacrylates, or polyacrylates.

17 Using zinc, titanium, or aluminium salts.

The woman who invented the dishwasher

Josephine Garis Cochrane is usually credited with inventing the modern dishwasher. In 1886, she patented a manual dishwashing machine that worked by spraying water. (Rudimentary dishwashers had been patented before, but their cleaning was done with mechanical rubbing, not water pressure.)

Josephine Garis was born in Ohio in 1839. At nineteen, she met and married William Cochran, a wealthy businessman, taking his surname but adding a final "e." The couple lived in Shelbyville, Illinois, for many years, leading a lively social life that involved organizing many dinners and parties. Apparently, Josephine was putting away the family china after a dinner when she noticed that some pieces had been chipped by her servants. Annoyed, she decided that from then on, she herself would wash the dinnerware. She soon found out that washing dishes was tiring and convinced herself that it was a complete waste of her time. It was an activity suited for a machine, but no such appliance was available. That was the moment, she would later say in many interviews, when she decided to invent and build one—despite never having received a formal education in mechanical engineering. She decided to mount dishes on a rack inside a copper container and spray hot, soapy water on them using water pressure generated by a manual pump. Then, more hot water rinsed and dried the plates and glasses.

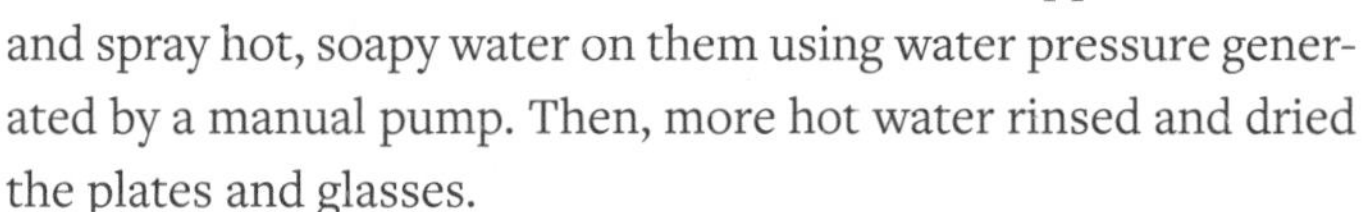

Her husband died in 1883, before she had a working prototype. With the money he left her, she employed a mechanic, and in 1885, she applied for a patent, obtaining it the following year and finally building a proper prototype. By 1888, the Garis-Cochrane company offered two differently sized models. The larger machine could wash and dry 240 items in about two minutes at full speed. These early dishwashers met with minimal success in private homes, partly due to their high costs. However, they were very popular in many large hotels and restaurants, where they saved on personnel costs and improved the cleanliness and sterilization of the dishes. The Garis-Cochrane appliance's ability to sterilize with boiling water persuaded hospitals and other institutions to install it in order to reduce the spread of germs, but this dishwasher never managed to conquer the household market.

After Josephine's death in 1913, her company continued manufacturing dishwashers until 1926, when it was absorbed by the Hobart Manufacturing Company, which later changed its name to KitchenAid (now owned by Whirlpool Corporation). It wasn't until 1949 that the dishwashers invented by Josephine Garis Cochrane became an increasingly widely used household appliance.

HAND-WASHING DISHES

Hand-washing methods

Different people from different cultures wash the dishes in different ways, and in recent years, this has become the subject of consumer behavior studies that carefully measure water, energy, and detergent consumption—sometimes even involving observations of individual behavior through cameras. These behaviors range from hand-washing dishes using very little water to letting the hot water run until all the washing and rinsing is done. One study that looked at resource consumption for twelve place settings worth of dishes showed that more water was always consumed when washing dishes by hand than when using the dishwasher, with the average water consumption varying from about 9 to 42 gallons (or 35 to 160 liters).[18] In another study (also based on twelve place settings), water consumption varied from about 5 to 124.5 gallons (or 18 to 472 liters).[19] Participants who washed dishes under running water used more than those who filled up their sinks, proving that there are more and less efficient ways to wash dishes by hand.

Hand-washing dishes is actually responsible for 50 percent of the water that comes from our home faucets.[20] While one dishwasher cycle uses about 3.5 to 4 gallons (or 13 to 15 liters), hand-washing requires an average of about 3 gallons (or 11 liters) per person per day.[21] Say we wash the dishes at least five

18 Petra Berkholz et al. "Manual dishwashing habits: an empirical analysis of UK consumers." *International Journal of Consumer Studies* 34, no. 2 (February 2010): 235–42.

19 Petra Berkholz, Verena Kobersky, and Rainer Stamminger. "Comparative analysis of global consumer behaviour in the context of different manual dishwashing methods." *International Journal of Consumer Studies* 37, no. 1 (January 2013): 46–58.

20 Christian Paul Richter. "Usage of dishwashers: Observation of consumer habits in the domestic environment." *International Journal of Consumer Studies* 35, no. 2 (March 2011): 180–6.

21 Rainer Stamminger, Angelika Schmitz, and Ina Hook. "Why consumers in Europe do not use energy efficient automatic dishwashers to clean their dishes?" *Energy Efficiency* 12, no. 3 (March 2018): 567–83.

times a week—it's easy to calculate the enormous quantity of water we could save by using more efficient techniques or a dishwasher. However, these studies also show that it's very difficult to change deep-rooted cultural behaviors and persuade people that we can wash the dishes in a different, more resource-efficient way.

People wash dishes by hand using at least three different methods. We can differentiate these based on where they put the dish soap.

DIRECTLY ONTO THE DISHES

Spread a little detergent directly onto the dish or onto a sponge, add a little water, and use mechanical action to clean the dishes, rinsing at the end.

IN A CONTAINER

Prepare a washing bowl containing dish soap diluted in water. Dip a sponge into this bowl occasionally, squeeze it out, and use it to clean the dishes, rinsing at the end. Here, the detergent is less concentrated than it is in the direct method.

IN THE SINK

Fill the sink or another large receptacle with water. Add a little soap, put the dishes in one at a time or all together, and wash them. Particularly dirty dishes can be left to soak for a few minutes to let the soap do its job and make dirt removal easier later. Lastly, rinse the dishes with clean water. This method dilutes the dish soap the most.

DID YOU KNOW?

If you've cooked pasta (or some other food that involves boiling water) and ended up with a particularly dirty pot or pan covered in caked-on residue, you can begin removing this residue using the hot pasta water and a little dish soap. The heat melts the grease, and the starchy water and surfactants start removing it. Then, you can wash the dish by hand or put it in the dishwasher.

Your preferred washing method depends on habit, but it's also a cultural question that varies from country to country. In Germany, most people who wash the dishes by hand do so using the sink method, and this is a widespread practice in Italy as well. Almost no one does this in Japan—instead, most people use the direct method. In Mexico, it's very common to use a separate container. In the US, the direct and sink methods are typical, while in Spain, all three processes are used.

Mechanical action is important when cleaning. All too often, we expect chemistry to do the work alone—for example, when we leave a dirty item to soak and get annoyed when it's still crusted over with burnt food. But if something is very dirty, it usually won't get clean without a lot of repeated scrubbing with a sponge, cloth, or scrubber.

THE MOST ECO-FRIENDLY WAY TO HAND-WASH

Okay, so you can't or don't want to use a dishwasher. You'll want to keep in mind that there are better and worse ways to wash dishes by hand when it comes to environmental sustainability and resource-saving. The best strategy for minimizing your use of water, detergent, and energy while maintaining a high level of hygiene involves two sink basins filled with water.[22] If you have only a single sink, you can use a large bowl as the second receptacle. Avoid washing dishes under running water at all costs: It's a pointless waste of resources.

1 Half-fill the first sink basin with hot water and the second with cold water. The former is for soaking and washing the dishes, and the latter for rinsing them.

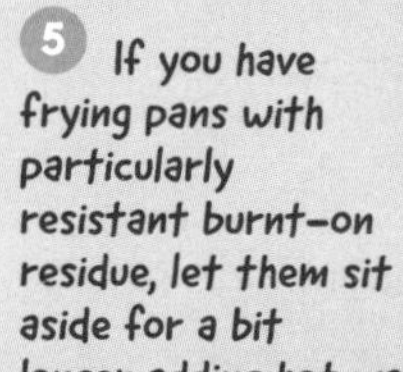

2 Add the soap, following the recommended dosage on the label. People often wrongly believe that more is better, but in fact, it's sometimes worse, since it forces you to use more water for rinsing.

3 While adding the soap to the water, try not to create unnecessary foam. Despite years of ads trying to convince us that the quantity of foam is important, it isn't. In fact, too much foam traps dirt rather than letting it sink to the bottom of the basin.

4 Throw all food residue in the trash, then soak the dishes. Start washing the cleaner plates first, since the dirtiest ones need more time to soak.

5 If you have frying pans with particularly resistant burnt-on residue, let them sit aside for a bit longer, adding hot water and a little dish soap.

6 Wash the dishes using a sponge or other cleaning tool. Rinse them in the second container, then place them in a dish rack to dry. Don't dry them with a dish towel, which is often a source of bacterial contamination.

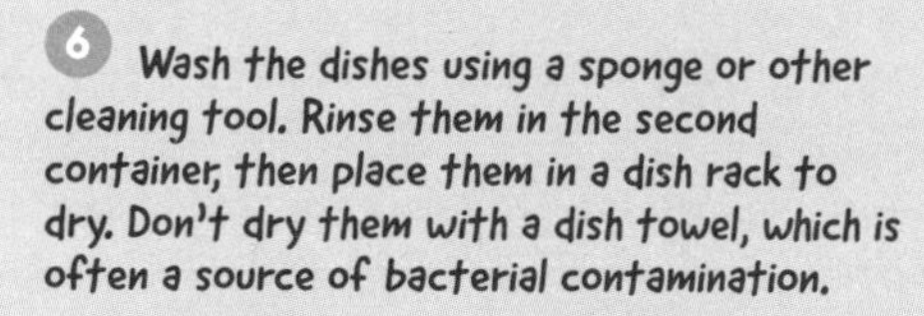

7 If you absolutely must dry the dishes with a dish towel, let them drain well first. Put your dish towels and sponges in the washing machine on a hot setting (140°F or 60°C) at least once a week, adding a little bleach to kill germs.

8 If the water in a container gets especially dirty while you're in the middle of washing, replace it.

22 Natalie Fuss et al. "Are resource savings in manual dishwashing possible? Consumers applying Best Practice Tips." *International Journal of Consumer Studies* 35, no. 2 (February 2011): 194-200.

DISH SOAPS

Dish soaps are versatile products for cleaning everything we use to eat and drink, from plates, pots, pans, glasses, and silverware to potato peelers, immersion blenders, polyethylene cutting boards, and baking trays. Although we call them soaps, like the products we use in the dishwasher, they're actually detergents. And like dishwashing detergents, their surfactant molecules remove dirt from dishes. Dish soaps are typically largely made up of surfactants (which can be 10 to 50 percent of their total content), usually anionic, nonionic, or amphoteric. They are also formulated to create a lasting foam without being too harsh on our skin. Beyond dishes, we can use them to wash our hands as well as kitchen surfaces, vases, and other home furnishings.

AT THE GROCERY STORE

Dish soap varieties

While dishwasher detergent comes in multiple forms, dish soaps all look pretty much the same: transparent plastic containers with a thick, colorful liquid inside. And while dishwasher detergents contain surfactants, enzymes, alkaline substances, calcium sequestrants, and other ingredients, dish soaps are made up of mostly water with one or more surfactants. Enzymes and sequestrants are much less common in these products. However, fragrances vary enormously: As well as the classic lemon or orange, there are the trendier lime, bergamot, and grapefruit, while those who don't like citrus scents can choose from mint, kiwi, pomegranate, peony, green tea, or vinegar. I've already mentioned that I prefer scent-free products, especially when it's something that comes into contact with skin, because some fragrances can be irritating if you have allergies. And I also just don't like added smells—although unfortunately, they're somewhat unavoidable because they're often used to cover the odors of other ingredients.

In any case, the biggest difference between dish soap brands is not fragrance or color, but the types and concentrations of surfactants they contain. This determines how many dishes can be washed using the same amount of soap. The magazine *Altroconsumo*, which compared different dishwasher detergents, also assessed sixteen dish soaps in 2021 based on the same criteria.[23] The results revealed surprising differences: The best-performing product washed thirty-seven dishes, while the lowest-ranked only washed fifteen. It's interesting to note that in the overall ranking, three of the products in the top four positions can legitimately be described as environmentally friendly. It's also notable that price has practically nothing to do with performance: The most expensive product (an organic detergent) finished tenth and costs almost three times as much as the winner, while the last-place product costs more than any of the top five and performed worse when it came to washing. However, the three cheapest products were ranked in eleventh place and lower. Finally, it's intriguing that two soaps in the top four are generic products marketed under grocery chain brand names.

---

23 "Detergents: how much do they impact on the environment? Discover this with the European CLEAN project," *Altroconsumo*, last modified March 8, 2021, altroconsumo.it/clean.

PRODUCT SPOTLIGHT: NELSEN LEMON

Let's see what Henkel's Nelsen Lemon dish soap contains.[24]

Can I make it at home?

No, because you wouldn't know where to get the surfactant (and pasta water definitely isn't a good substitute). But dish soap is often used as an ingredient in other household products thanks to its surfactants, which are selected to clean a wide variety of dirt.

INGREDIENT	FUNCTION
Water	Solvent
Sodium lauryl sulfate and amidopropyl betaine	Anionic and amphoteric surfactants
Sodium chloride	Viscosity regulator
Limonene and citral	Fragrances
Methylisothiazolinone and benzylisothiazolinone	Preservatives
Subtilisin	Protease enzyme that breaks down proteins
Acid Yellow 23	Colorant
Denatonium benzoate	Bitter flavor

24 Henkel product ingredients can be viewed at mysds.henkel.com.

IS FOAM IMPORTANT?

In general, foam is not necessary for hand-washing dishes (or cleaning anything else, really). However, consumers still tend to associate this characteristic with product effectiveness. Try washing your hair with a shampoo that produces very little foam: You'll probably get the (not necessarily correct) feeling that it's not cleaning your hair properly. This is why dish soap formulas include added surfactants that produce extra foam and help the foam stick around longer.

The idea that lather is important to cleaning may have historical origins. As we've seen, back when people used actual soaps, hard water progressively reduced those soaps' effectiveness. Calcium and magnesium salts literally remove the surfactants from the soap, and the absence of foam shows that these surfactants have become ineffective. To remedy the problem, people needed to use larger quantities of soap. This is no longer an issue for modern dish "soaps," although the quantity of foam a dish soap produces may help you figure out how to properly dose it.

DID YOU KNOW?

Why are dish soaps so thick? Higher viscosity makes dispensing and dosing easier—but also, consumers associate this quality with higher concentration and, therefore, greater effectiveness. Special substances are added to make dish soap thicker. But since the soap needs to dissolve quickly once it's poured into water, other ingredients must be included to facilitate dilution. Sometimes it's enough to put a little sodium chloride (common table salt) into the formula to increase viscosity.

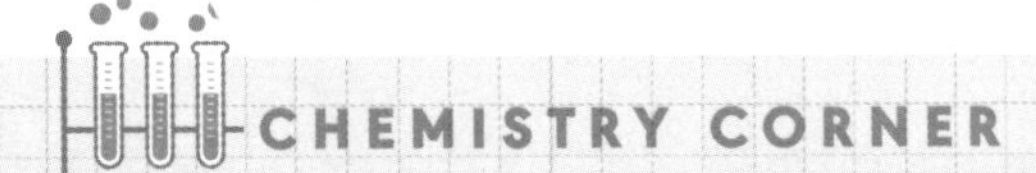

CHEMISTRY CORNER

Protecting your hands

If you wash dishes by hand and hate wearing gloves, you'll want to look for dish soap that is kind to your skin. Otherwise, you have to worry about dryness, redness, inflammation, and swelling. In general, nonionic surfactants are gentler than anionic ones, and amphoteric surfactants are even better.

10

Disinfectants

Washing and disinfecting are not the same thing. I'm sure some of you are aware of this, but you may be in the minority: One poll from 2020 found that over 15 percent of Europeans mix up these two practices and see them as equivalent.[1] Disinfecting is also not the same as deep cleaning, while sterilizing means something else altogether . . . this is where things get a little foggy.

The terminology in this area can cause confusion when we're at the grocery store looking for something to disinfect our toilet. Product labels don't help, and they may even lead us further astray: What on Earth does it mean when a certain cleaner says it "removes" 99 percent of bacteria? Does it kill them? But it would be too simple to put all the blame on advertisers and manufacturers—some of it certainly belongs to the complex legislation regulating disinfectants, which has to take into account antimicrobial efficacy, user safety, environmental impact, consumer communications, and much more.

1 AISE (International Association for Soaps, Detergents and Maintenance Products) and IFH (International Scientific Forum on Home Hygiene), *Developing household hygiene to meet 21st century needs: A collaborative industry/academia report on cleaning and disinfection in homes & Analysis of European consumers' hygiene beliefs and behaviour in 2020* (Brussels: April 2021).

WHAT IS DISINFECTING?

Think back to the last time you used a disinfectant, and tell me the truth: Did you read the instructions? Did you measure the correct dose? Did you dilute it as required? Or did you eyeball it, judging how much you were using by the number of "glug-glug-glug" noises it made coming out of the container? If accurate dosing matters when using laundry or dishwasher detergent, it's even more important for disinfectants, since your and your loved ones' health may depend on it. Ask yourself why you bought a disinfectant for household cleaning: Do you really need to kill microorganisms, or do you simply want to maintain good hygiene in your home?

Any dictionary will tell you that the term "hygiene" covers all the practices we use on our bodies, possessions, and living environments to maintain good health. We restore normal hygienic conditions with methods referred to as "deep cleaning," which reduce the number of bacteria present to a level that poses no risk to us.

Washing your hands before eating and after going to the bathroom is about hygiene, and so is disinfecting the cutting board you've just used to fillet a raw chicken breast. However, putting a shirt that you spilled sauce on in the washing machine is less about hygiene and more about getting rid of dirt, as is dusting your furniture. There is, of course, a degree of overlap between the two, since getting rid of dirt is a necessary condition for achieving an adequate level of hygiene. But it's important to be clear about the differences so we can avoid underestimating the dangers of specific contaminants and neglecting practices that could spare us from potential health problems—or using disinfectants for no good reason.

What is a disinfectant?

We're trying to be precise about the definitions of the words we're using, but it doesn't help that in our everyday language, certain terms often pick up meanings that diverge from the national and international regulations covering the world of cleaning. We commonly say "disinfectant" when talking about a chemical agent that kills or deactivates microorganisms, and when we refer to "disinfecting," we mean using that product against microorganisms.[2] Seems simple, but let's look at why it isn't—starting with those definitions, which need refining.

You can easily find substances that are capable of killing at least some bacterial species just by walking around the house. As one example, if you took the gloss left over from repainting a wrought iron gate and applied it to bacteria, it would probably kill some of them. The organic solvent you used to dilute the gloss would probably do the same thing—but you would never disinfect your toilet bowl with either of them. Vinegar must be able to kill some bacteria (and, in certain

2 "Microbes" or "microorganisms" refer to any microbiological entity (cellular or not) that's capable of self-replicating or transferring genetic material. This includes microfungi, viruses, bacteria, yeasts, molds, algae, protozoa, worms, and microscopic parasites.

quantities, stop them from multiplying), since it's been used as a food preservative for centuries. But if you think about it, the acetic acid in vinegar is literally made by bacteria, as is ethyl alcohol, which can also preserve food and drinks. And as we learned during the COVID-19 pandemic, while ethyl alcohol can destroy the virus responsible for this disease, it only does so when used in specific ways and concentrations—washing hands or surfaces with vinegar doesn't work. In short, even the idea of "killing microorganisms" is complicated.

According to the US Department of Health and Human Services, disinfectants fall into a category of substances called "biocides" (which also includes preservatives, antiseptics, fungicides, and insecticides).[3] The term "biocide" encompasses a vast number of products intended to kill, eliminate, deactivate, or disarm dangerous organisms such as bacteria, viruses, insects, and other creatures. The word is composed of the prefix "bio-," indicating a living organism, and suffix "-cide," which means "capable of killing"—even though a biocidal product doesn't necessarily kill its target organism.

Some substances are bactericides: They kill bacteria by destroying these cells through various chemical and biochemical mechanisms. Others are only bacteriostatic agents, meaning that rather than killing, they prevent bacteria from multiplying—for this reason, they're used as preservatives. Then we have sporicides, which are effective against spores, particular structures developed by some bacteria in adverse environmental conditions. Bacterial spores are one of the dormant forms of bacteria, surviving by exhibiting minimal respiration and metabolism as well as reduced enzyme production while they wait for better conditions (for example, a more favorable pH value). Spores are extremely resistant not only to many biocides but also to physical means of eradication, like heat.

So, the term "biocide" is a big umbrella that covers varied products that are designed to target different needs. Even if the leftover gloss for our wrought iron gate is technically

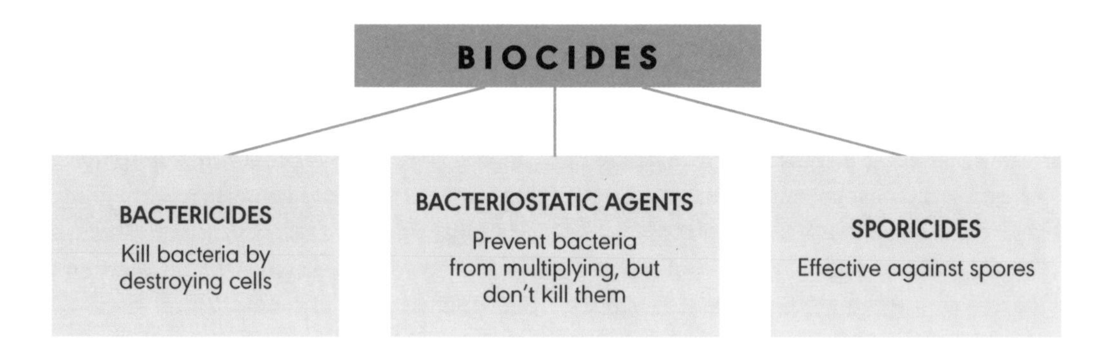

3 "Biocides & Potential Respiratory Health Outcomes," *National Toxicology Program*, last modified March 7, 2023, ntp.niehs.nih.gov/whatwestudy/assessments/noncancer/ongoing/biocides.

a biocide, that doesn't mean it can actually be sold as one.

To kill bacteria, viruses, and other microorganisms (or at least stop them in their tracks), disinfectants and other biocides must act against one of their biological parts—perhaps the membrane or cell wall, a particular enzyme, or really any structure that's essential to a microorganism's functions. However, these structures can also be present in other living beings, humans included, so once disinfectants are released into the environment, they can prove toxic to the organisms they come into contact with. It's for this reason that government organizations strictly regulate this category of substances.

Only disinfect when necessary

When we know the differences between cleaning and disinfecting, we can figure out which option is best for a given situation. We all use disinfectants at home, but are we sure we're using them thoughtfully? I don't think anyone regularly disinfects their shelves full of knickknacks, and I believe that you're probably mostly using disinfectants in your bathroom and kitchen. Good—if this is true, you're making a sensible evaluation of benefit versus risk.

I am, however, pretty sure that at least some of you also use a disinfectant to wash your living room floor. Why? Think about it for a moment. Bacteria are everywhere, all the time—and some of them are even pathogens, which can make us sick. But how likely are you to be infected by a pathogenic bacterium lurking on your living room floor? Do you eat pizza on that floor? If you dropped a little sugar, would you lick it off the tiles? You can disinfect that floor as often as you want, but the microorganisms will likely come back before the day is over.

Instead, I recommend concentrating on the areas that pose the greatest actual risk. (As we discussed in chapter 1, the difference between risks and hazards is crucial.) Of course, you may have a pet running around, a small child crawling all over the place and putting their hands in their mouth, or a household member who's immunocompromised—and in these cases, the risk is higher. Otherwise, we need to let go of the attitude best described by a riff on a famous quote: "I love the smell of bleach in the morning. . . . It smells like victory!"[4] Disinfecting every floor and wall in the house is simply excessive in most cases, and even kitchen and bathroom floors present a smaller risk than most people think.

The problem is that while standing on that immaculate kitchen floor, we often engage in high-risk behaviors like washing chicken breast, or cutting a cooked roast on the same board we used when the meat was raw. We don't regularly disinfect kitchen sponges and dishcloths, and we may not carefully wash our hands with soap after using the toilet (no, water alone is not enough).

Hygiene is about health, and so if we want to use the right products exactly when they're needed, we first need to understand how we get infected by bacteria and learn to take appropriate precautions.

4 Robert Duvall was actually talking about the smell of napalm. *Apocalypse Now*, directed by Francis Ford Coppola (Omni Zoetrope, 1979).

FIGHTING DISEASE

As I wrote this book, we were still going through the COVID-19 pandemic, so the "chain of infection" was a familiar concept: The microbes responsible for communicable diseases spread from one living being (human or animal) to another through a particular medium. That may be air (as in the case of SARS-CoV-2) or bodily fluids (for a virus like HIV). Contact between people may be direct, or it can happen indirectly through an object they both interact with. In short, infection can occur in a number of ways, and if we're infected, we may infect someone else in turn. We need to break this chain by using the means most appropriate to the situation, whether it's a face mask, a disinfectant, or a washing machine.

DID YOU KNOW?

When we talk about bacteria, or more generally about pathogenic microorganisms, we're referring to microbes capable of causing disease (or "pathology") in humans. While those particular microbes must be kept at bay, they're only a small fraction of all microbes in existence, most of which carry out important functions for us and our environment. That's why we shouldn't obsess about microorganisms: There's no reason to eliminate them from our lives.

WARNING! Washing chicken in the sink is a common practice, but it's wrong—stop doing it! It risks contaminating the sink and everything nearby with all kinds of pathogenic bacteria, including *Salmonella*, *E. coli*, and *Campylobacter*. (To remove these bacteria, chicken meat must be cooked, not washed, and of course, it should never be consumed raw.) The reckless decision to clean chicken in the sink actually increases the risk of infection, since water inevitably splashes the surrounding surfaces, maybe getting all over that apple you'll eat later. If you're squeamish about the slime on the surface of your poultry, just remember that it comes from proteins that will denature and coagulate during the cooking process.

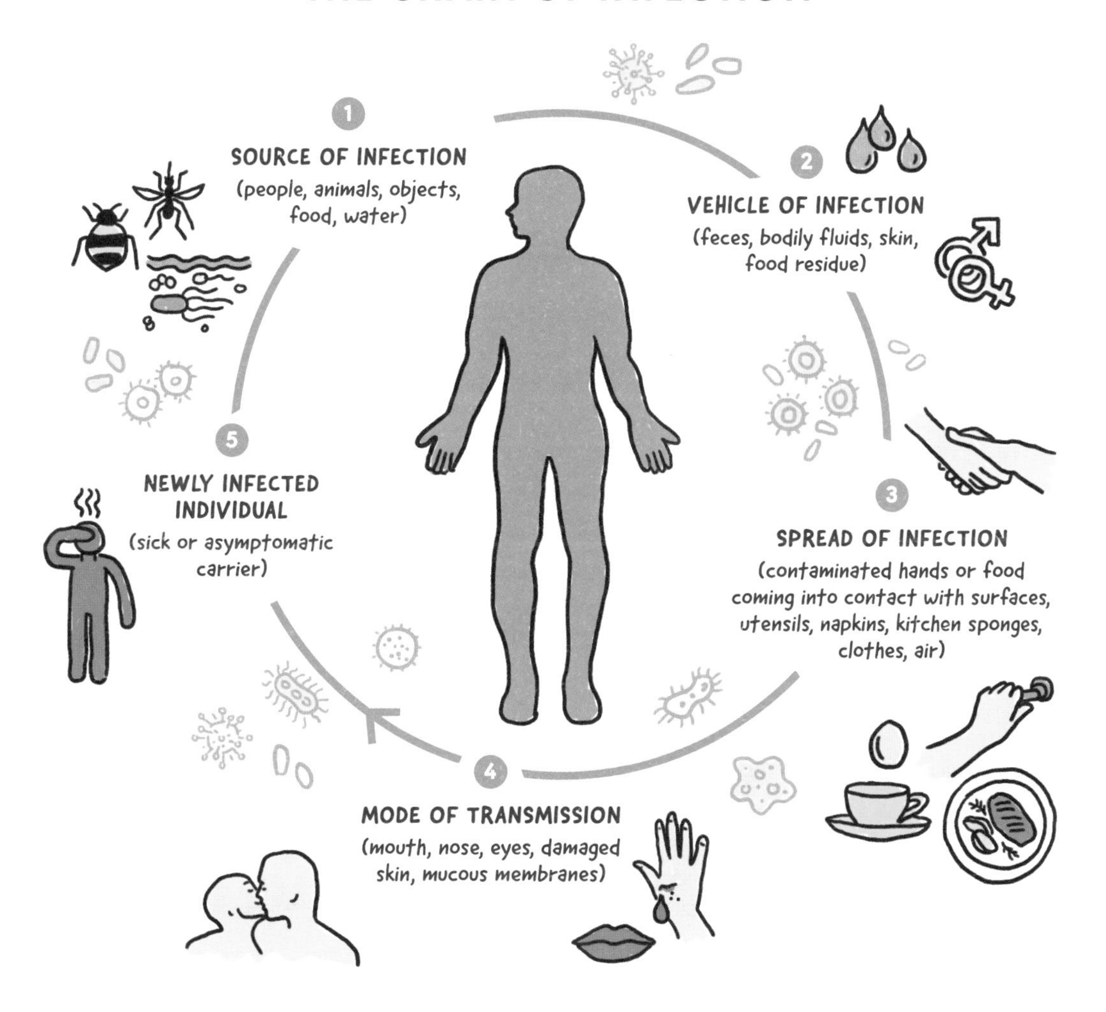

Transmission risks of different surfaces

Once we know how a pathogen is transmitted from one living being to another, we can identify situations that create higher risk of transmission.

Have you noticed a recurring theme in what you've read so far in this chapter? If you guessed "hands," you're correct. Hands are both a vehicle of self-infection through touching our own bodies and a source of contamination for objects in our environment. This is why washing (and sometimes disinfecting) our hands whenever they may have become dirty is important.

The other key surfaces in our living spaces are cutting boards used for raw food, toilets, dish towels and sponges, waste, walls, floors,

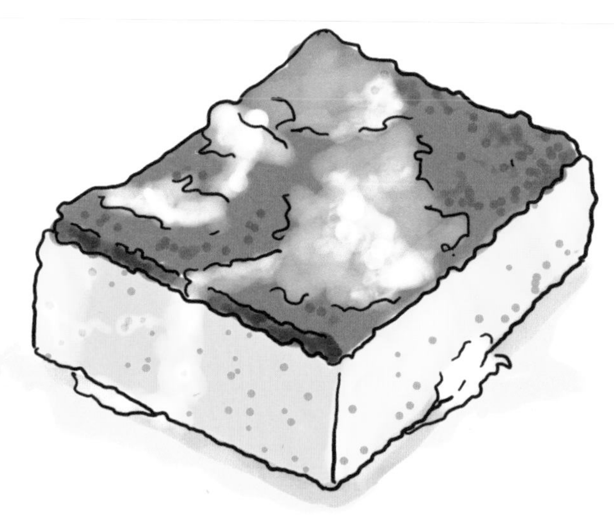

furniture, and any objects that multiple people frequently touch.[5] Obviously, individuals should think about what's most likely to get contaminated in their own lives, but we can all develop a general ability to identify areas and items associated with varying degrees of risk. (I'm sure that you wash your hands much more often than the floor, and that you change and wash your underwear more often than you wipe down your shower door.)

DID YOU KNOW?

According to the so-called five-second rule, when a piece of food falls on the floor, you can still eat it as long as you pick it up within five seconds. This rule assumes that rescuing your snack fast enough means there's no time for it to become contaminated—but from a scientific point of view, that makes no sense whatsoever. Bacteria, viruses, and fungi don't take five seconds to hop onto a fallen piece of bread using their tiny little (nonexistent) legs. Actually, they adhere immediately on contact with the morsel.

While hands are our biggest concern, we also need to worry about things that often come into contact with or contain large quantities of microorganisms: for example, surfaces that touch raw food (especially animal products), kitchen sponges, utensils, and anything that we frequently interact with using our bare hands. Much of this is common sense, though—I'm reasonably sure that if you handle raw fish, you wash your hands thoroughly with soap before touching the TV remote or your cell phone.

Surfaces that are generally less risky include clothes, bedsheets, towels, sinks, and toilet exteriors. Clothes and bedsheets present lower transmission risks than dish towels, although textiles that come into direct contact with our skin are a bit more problematic. In any case, it's unnecessary to disinfect any of these fabrics under normal conditions. A regular washing machine cycle will do, although if you're washing at low temperatures, you should use a deep cleaner like activated percarbonate or delicate bleach.

Like walls and furniture, the floors in our homes rarely pose a significant risk of pathogen transmission—despite some people's obsession, encouraged by cleaning product commercials, with getting them "clean enough to eat off." If you touch the floor, your hands can become a vehicle of contamination, but the best solution is washing your hands, not disinfecting your linoleum.

5 Objects that, if contaminated, can transmit a communicable disease to a new host are known as "fomites."

HIGHEST-RISK HOME ACTIVITIES[6]

1 Handling food

2 Eating with your hands

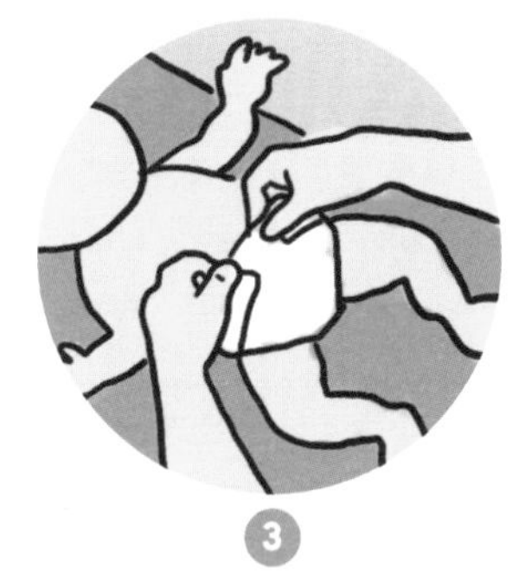

3 Using the toilet or changing a baby's diaper

4 Coughing, sneezing, or blowing your nose

5 Touching surfaces frequently touched by others

6 Handling or laundering clothing and household linens

7 Interacting with pets and other domestic animals

8 Disposing of domestic waste

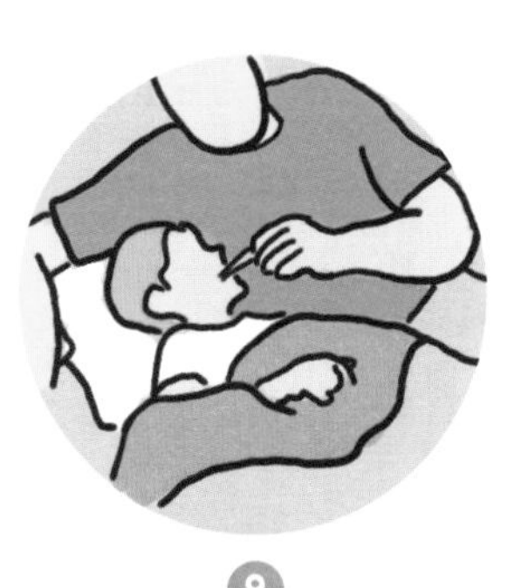

9 Taking care of a sick family member

6 Identified by the International Scientific Forum on Home Hygiene and Health (IFH).

You may be surprised not to find toilets in one of the highest-risk categories. But it's not all about the presence of microorganisms—toilets are made to avoid spreading bacteria and viruses around. It's most important to focus on the things that our bodies make direct contact with: the toilet lid, seat, and handle.

Cleaning the bathroom using a deep cleaner like bleach or sodium percarbonate or even a disinfectant as well as a detergent has the advantage of reducing any possible contact between your hands and pathogenic bacteria when you're wringing out cloths or handling a mop and bucket. However, it's important to point out that every time we flush the toilet, tiny water droplets get dispersed into the air, potentially carrying microorganisms several feet away. Closing the lid before flushing may reduce this diffusion, but not by much, because of the gap between the lid and the seat—although at least you're giving the resulting aerosol time to settle before you lift the lid again. Therefore, it's impossible to

TRANSMISSION RISK LEVELS FOR HOUSEHOLD SURFACES

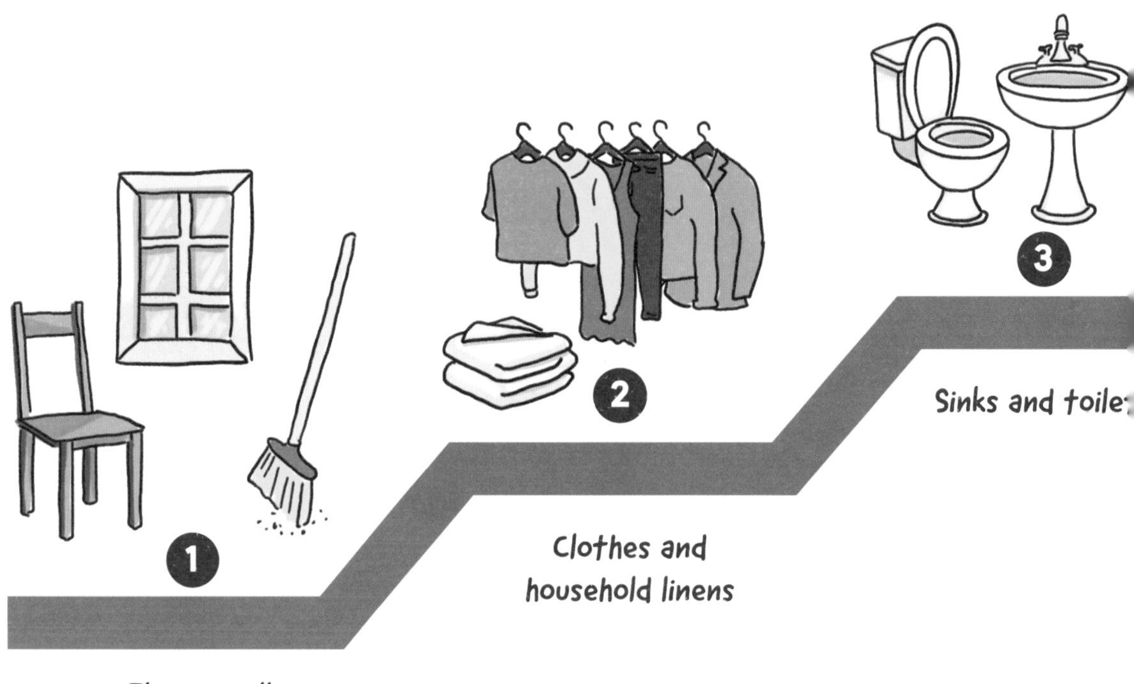

completely avoid touching microbes in this area.

If visualizing microorganisms floating around your bathroom makes your skin crawl, since you certainly can't disinfect the whole room after each flush, just consider this a good incentive to always wash your hands after using the toilet. You can also open a window for improved airflow. Obviously, you should never bring food into the bathroom—but even if you don't do that, you're probably scrolling on your phone or reading newspapers or comics. You likely leave those printed materials next to the toilet for a few hours or days before carrying them to other parts of the house . . . so think about what's quietly settling onto the pages.

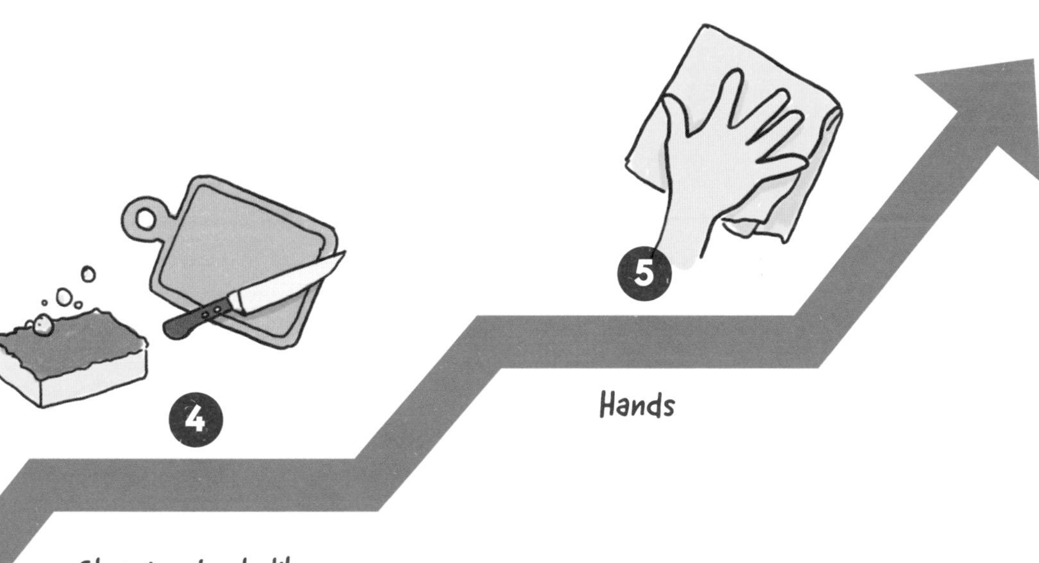

HOW DISINFECTANTS WORK

To paraphrase another fictional tough guy, when a microbe meets a disinfectant, that microbe will be a dead microbe.[7] In this situation, several things happen. (I'll focus on bacteria as my example here, only referring to other microbes like viruses and fungi when I need to highlight their specific characteristics.) First, the disinfectant interacts with the bacterium's surface and external biological structures. In some cases, the destructive action begins right there on the cell envelope, but more often, the disinfectant molecule penetrates that envelope to attack an internal target, whether that's an enzyme, DNA, or any other biological structure the bacterium can't live without.

Disinfectants can be divided into two groups: oxidizing and non-oxidizing. The latter group encompasses alcohols, phenols, and quaternary ammonium compounds. Oxidizing disinfectants include chlorine compounds, hydrogen peroxide, and iodine. They have a broader spectrum of activity and are also effective against spores, but at the same time, they have a longer list of contraindications because they pose greater risks to human health.

A disinfectant's effectiveness is determined by many factors.

Microbial load

I may be stating the obvious, but just to be clear: A disinfectant is more effective when it's used on a surface with a lower density of microbes.

Microbial diversity

A generic disinfectant is more effective on a pure microbial population than one that includes many different species. It's unlikely that a standard disinfectant can kill all the microbes in a diverse population—some will survive because different species have different levels of resistance to a particular product. Fungi tend to be more resistant than bacteria, and bacterial spores most resistant of all. The question of whether the survivors will be able to multiply and become a problem again comes down to time and the availability of nutrients.

Bacterial species

Gram-positive bacteria tend to be easier to eliminate than gram-negative bacteria. (Danish bacteriologist Hans Christian Joachim Gram developed this commonly used technique for classifying bacteria, which involves staining them to make them more clearly visible under a microscope.) However, some gram-positive bacteria can produce spores that are extremely resistant to disinfectants.

Biofilm

Bacteria suspended in water (perhaps loosened from a piece of fabric by laundry

7 Or as Clint Eastwood said, "When a man with a .45 meets a man with a rifle, the man with a pistol will be a dead man." *A Fistful of Dollars*, directed by Sergio Leone (Jolly Film, Constantin Film, and Ocean Films, 1964).

detergent) are easier to kill than those clinging to a surface. The worst imaginable situation, however, is when bacteria form a biofilm: a complex biological aggregation of microorganisms that cover a surface by secreting an adhesive protective matrix. This sort of "colony" is sometimes made up of several bacterial species, and it is very difficult to eliminate. Biofilm is estimated to be responsible for approximately 60 percent of microbial infections, since in this form, bacteria are better at resisting disinfectants, antibiotics, and attacks from a host's immune system. (Yes, biofilms can form in complex living organisms just as well as they do in a pipe or washing machine.)

Dirt

The presence of dirt particles like grease and dust can act as a physical barrier between the microbes and the disinfectant, reducing or canceling out the latter's effectiveness.

Surface type

The type of surface you're cleaning is hugely important. Stainless steel is hard and smooth with no irregularities, making dirt and microbe removal a lot easier than rougher surfaces like plastic or ceramic tiling. Also, bacteria can penetrate porous materials like wood, thereby avoiding contact with disinfectants. Some porous surfaces like cement or walls that haven't been properly treated can be a nightmare when you're trying to get rid of mold infestations. A disinfectant may never be able to reach fungi that have deeply penetrated into the pores, and in these extreme cases, removal and rebuilding may be the only solution.

Temperature

In most cases, the higher the temperature, the faster disinfectants kill. But in practice, this rarely matters, since we almost always disinfect at room temperature.

pH

pH is important because of its impact on disinfectant molecules' ability to adhere to bacteria's external membranes. Some products are more effective in acidic environments and others in basic environments.

Concentration

Here's another factor that may sound trivial but is actually crucial: Below a certain critical level of concentration, a disinfectant is no longer effective. This is why you should always read the instructions on the label! And don't think that "more must be better," because it isn't. A classic example of this is a substance some of us got to know during the pandemic: ethyl alcohol. The optimal ethyl alcohol concentration for eliminating SARS-CoV-2 is 70 percent, while concentrations below 60 percent or above 90 percent have little to no effect.

Contact time

A disinfectant takes time to do its work: Some are very fast, and others are slower. Contact time is the period a disinfectant requires to reach the desired level of microbial load reduction. You should also always read the instructions to confirm the minimum time that a product needs to be effective.

DISINFECTANT REQUIREMENTS

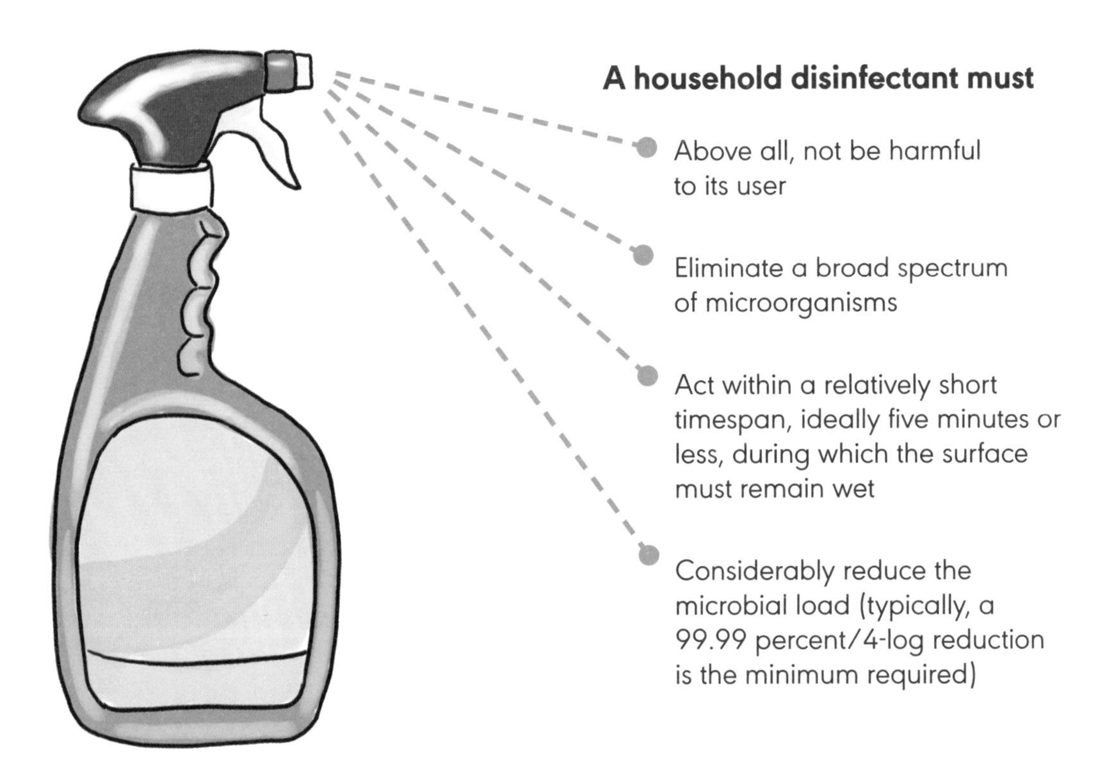

Log reduction

Disinfectant effectiveness is measured in terms of "logarithmic reduction," which quantifies how many times initial contaminant concentration is divided by ten when a product is used. For example, in one test, 10^7 colony-forming units (CFUs) are inoculated with a particular bacterium. After disinfectant is applied, 10^5 CFUs are left. The reduction factor is 100, which is 10 times 10 or "2-log." The log reduction required for a product to be considered effective depends on your country's regulations. Some ask for a 4-log reduction (which eliminates 99.99 percent of microorganisms).

Different forms of disinfectants

There are so many disinfectants in the household products aisle (and this book doesn't even touch on those formulated exclusively for health care purposes, like disinfecting wounds). Their forms and objectives are varied enough that it's hard to split them into categories—but I'll try to do just that by grouping them according to the way they're used.

CONCENTRATES

Some products have to be diluted in water because they're sold at higher concentrations than they're meant to be used at. These include disinfectants based on sodium hypochlorite, which some doctors recommend to pregnant women for eliminating bacteria from fruit and vegetables. (I'll take this opportunity to remind you that neither chlorine-based products nor the ever-present baking soda have any effect against the parasite that causes toxoplasmosis. You can read more about toxoplasmosis risk in the next chapter.)

Although these products can be cost-effective, failing to read the label and using a concentrated substance without diluting it first can lead you to use more than necessary, and some of these disinfectants are harmful in their undiluted form. For example, household bleach (which contains sodium hypochlorite but is not marketed as a disinfectant) is commonly sold in concentrations between 5 and 9 percent, but sodium hypochlorite acts as a disinfectant at just 0.1 percent and is very corrosive when undiluted. That's why bleach includes instructions for adding it to the washing machine that take into account the machine's capacity—that is, the amount of water that will dilute the bleach.

SPRAYS

Spray-on disinfectants can present problems because if the dispenser creates a mist that's too diffuse, the substance can get dispersed into the air as well as onto the surface you're trying to clean. This leads to both reduced effectiveness and a danger to anyone in the vicinity who breathes in the disinfectant.

CLOTHS AND WIPES

These come pre-saturated, typically with alcohol. The idea is that the product helps remove dirt as it disinfects. Therefore, the material must be filled with enough disinfectant to wet the whole surface and stick around for the ideal contact time.

As convenient as cloths and wipes are, studies have shown that they're less effective for cleaning than wetting a surface before wiping it with a dry cloth. This may be because these products don't always contain enough disinfecting liquid to cover the entire surface, or because the fact that they're already saturated prevents them from reabsorbing liquid, running the risk of spreading rather than removing dirt.

GELS

Gel products are liquids with added substances (polymers, usually) that increase their viscosity, helping them cling to slanted and vertical surfaces for longer. These are typically used to clean the inside of your toilet.

WASH BEFORE YOU DISINFECT

I've already said it, but I'm going to reiterate it: Washing is not the same as disinfecting. And the opposite is also true: A disinfectant does not wash. For example, we know that bleach alone doesn't wash, since sodium hypochlorite is not a surfactant (that is, it can't remove dirt).[8] Sodium hydroxide may lend a little detergent power with its alkaline pH, but if you cleaned the floor with bleach alone, you'd risk spreading dirt all over the tiles.

Not only are people who clean the bathroom floor with a little diluted bleach not really washing it (other than with the water added for dilution), but they might not even be disinfecting it. Most disinfectants are effective when used with or after detergents, so you need to remove the dirt first. This is because in order to do its work, a disinfectant molecule needs to come into contact with a microbe, and if said microbe is hiding under a dust particle or grease stain, the disinfectant can't reach it. Also, disinfectant molecules may react with the organic matter present in the dirt, which reduces their disinfecting power. Sodium hypochlorite, for instance, is a very effective disinfectant, but it oxidates all organic matter in its way. If it encounters too much dirt, there won't be enough of it left to kill the bacteria you unleashed it on.

Dirty surface

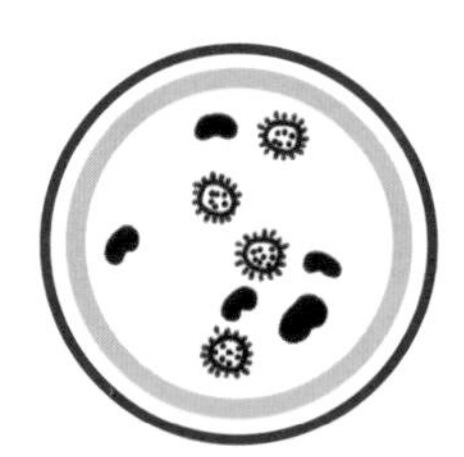

Surface after washing

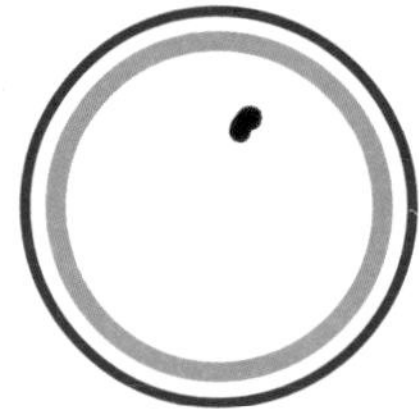

Surface after washing and disinfecting

Washing also means physically removing microorganisms. Next time you're at the grocery store, have a look at the various cleaning products that claim to have some effect on microbes. Some are labeled as disinfectants or biocides, but others are advertised as "removing bacteria," which is actually what detergents do: loosen bacteria and viruses from a surface, then carry them away. Handwashing also protects us from infectious diseases in just this way—some soaps and detergents may have the added benefit of killing a certain quantity of microbes, but that's not their main task. Overall, you should always wash before you disinfect so that there are fewer microorganisms for the disinfectant to eliminate.

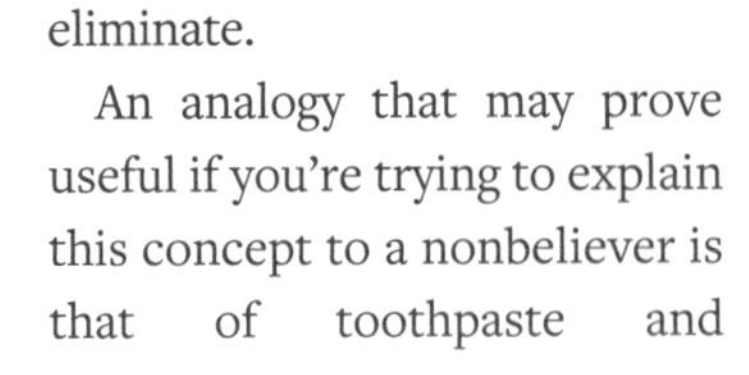

An analogy that may prove useful if you're trying to explain this concept to a nonbeliever is that of toothpaste and

8 However, there are products on the market that contain a detergent as well as all the ingredients commonly found in household bleach.

mouthwash. Toothpaste is formulated to clean teeth, and its effect mostly comes from the toothbrush's mechanical action, which loosens bacteria and allows them to get washed away when you spit. After you brush, you use mouthwash, which includes an ingredient that's capable of reducing bacterial load. But your dentist would hardly be pleased if you told them that you stopped using toothpaste when you started using mouthwash. The same goes for household surfaces—and too often, I've noticed a habit of using disinfectants in excessive quantities while forgetting that ordinary cleaning does the bulk of the work, paving the way for biocides.

STERILIZATION

Sterilization allows us to reach the highest degree of microorganism elimination. This procedure entails completely destroying bacterial and fungal spores as well as all bacteria, viruses, and other organic microorganisms. You may be familiar with this idea if you've canned your own preserves, but in fact, proper sterilization isn't really something you can do at home. Boiling the jars and rigorously following the canning procedure undoubtedly makes your homemade products safe to eat—just don't think of them as sterile. For just one example of microbes that could be sticking around, bear in mind that the spores of *Clostridium botulinum* type A (which cause botulism) can resist being boiled at 212°F (100°C) for up to six hours.

DID YOU KNOW?

Washing your hands thoroughly with regular hand soap removes 99 percent of all microbes present.

PROPER HAND-WASHING TECHNIQUE

HOUSEHOLD DISINFECTANTS

Around the house, you'll probably be able to find at least one product with disinfectant properties—even though, as you're now aware, that doesn't mean it's legally classified as a disinfectant or will always act as one. Let's get to know these household substances better.

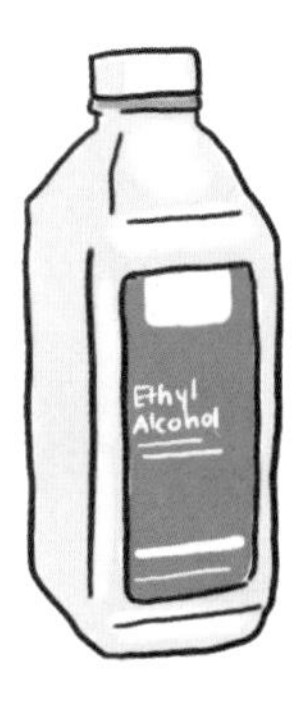

Ethyl alcohol

Alcohols have disinfectant properties that have long been exploited in various industries. These substances act on bacterial cells' external walls, making them permeable and thus triggering the denaturing of their proteins, leading to plasma leakage and eventually death.

You may have read online that ethyl alcohol is not a disinfectant, maybe even with a reference to an old scientific paper. Well, that's wrong—but the truth is that pure ethyl alcohol is not a particularly effective disinfectant. Alcohols are better at disinfecting when they're diluted with a little water, with 70 percent being the most effective dilution. (Isopropyl is the type of alcohol most commonly used in this concentration for industrial purposes.) Outside of this window of effectiveness, ethyl and other alcohols do little or nothing.

One drawback of ethyl alcohol is that it

WARNING! Sadly, household accidents caused by the ingestion of disinfectants and other cleaning products are far from rare. Children are most at risk, since they tend to be attracted to liquids and substances in brightly colored containers. All it takes is leaving just one bottle open, so make sure to always close containers properly. Some spray containers even have nozzles that can be shut off, which also prevents the product from evaporating between uses and protects people from getting sprayed by accident.

If someone swallows or inhales a potentially toxic substance, immediately call emergency services—you'll be put in touch with the nearest branch of Poison Control.

Here are two things you should never, ever do if a toxic substance is ingested:

1. Never induce vomiting unless you're explicitly advised to do this by Poison Control. It may seem like a good idea to get the substance out of the stomach, but it can actually prove counterproductive. Corrosive substances can do as much harm to mucous membranes on the way back up as they did going down, and if foam is produced, it may enter and damage the lungs.

2. Don't drink milk. Although this advice is sometimes handed out, milk is useless at best and may even compound the damage through its ability to emulsify some toxic substances.

evaporates very fast, which creates a risk of insufficient contact time with microbes. For this reason, classic alcohol-based hand sanitizers come in gel form, which not only is easier to spread on your hands, but also evaporates more slowly. Also, alcohols don't work against spores or certain viruses—for those, you'll need a stronger disinfectant.

If you've poured an alcohol-based disinfectant solution into a spray bottle, don't be stingy when applying it: You need to get the surface wet, since there's no disinfecting action without contact. Don't be impatient, either: Let the spray sit for at least thirty seconds before wiping the surface off with a cloth. The problem with spray bottles is that they dispense the product in lots of tiny droplets that don't uniformly wet the surface. To increase contact with the surface and the amount of product released, lay down a paper towel soaked in the alcohol solution after spraying.[9]

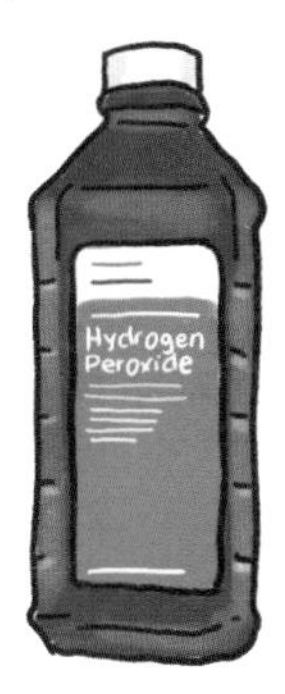

Hydrogen peroxide

You probably have a bottle of hydrogen peroxide in your medicine cabinet. It's usually sold at 3-percent concentration, but since it's a fairly unstable product, it's best to always check the expiration date—the actual concentration can lower over time, particularly if the bottle is left open or stored in a warm place. Ordinary 3-percent hydrogen peroxide is a broad-spectrum disinfectant that's active against viruses, bacteria, fungi, and bacterial spores (although it destroys the latter more effectively at higher concentrations). Nowadays, it's not recommended for direct use on cuts and scrapes because it seems to delay healing by destroying new cells. Since hydrogen peroxide degrades so rapidly, it's not hazardous to the environment. Concentrations of up to 10 percent can be obtained for industrial use, but they're not normally available over the counter since this higher-concentration variety is a dangerous oxidizer.

Sodium hypochlorite

Sodium hypochlorite is one of the oldest disinfectants, and it's effective against all bacteria (both gram-positive and gram-negative) as well as spores, viruses, and fungi. It's also one of the fastest disinfectants thanks to hypochlorous acid's ability to attack nitrogenous compounds such as amino acids, disintegrating their proteins. Since it works against spores, sodium hypochlorite is also found in products for disinfecting food (which should be done only when recommended by a doctor)—in this case, it's pre-diluted to concentrations between 0.1 and 0.2 percent or provided at a concentration of approximately 1 percent and then diluted in water before use.

Sodium hypochlorite is also present in household bleach at much higher concentrations, but these formulations must never be used to disinfect food (not even after dilution) because of the other substances they

9 This trick was suggested by Alessandro Mustazzolu, a microbiologist from the Italian National Institute of Health.

may contain. Household bleach is generally too concentrated to be used as is—it's a very corrosive product, and sadly, every year it's involved in multiple domestic accidents involving children who pour it on themselves or drink it from containers that have been left open.

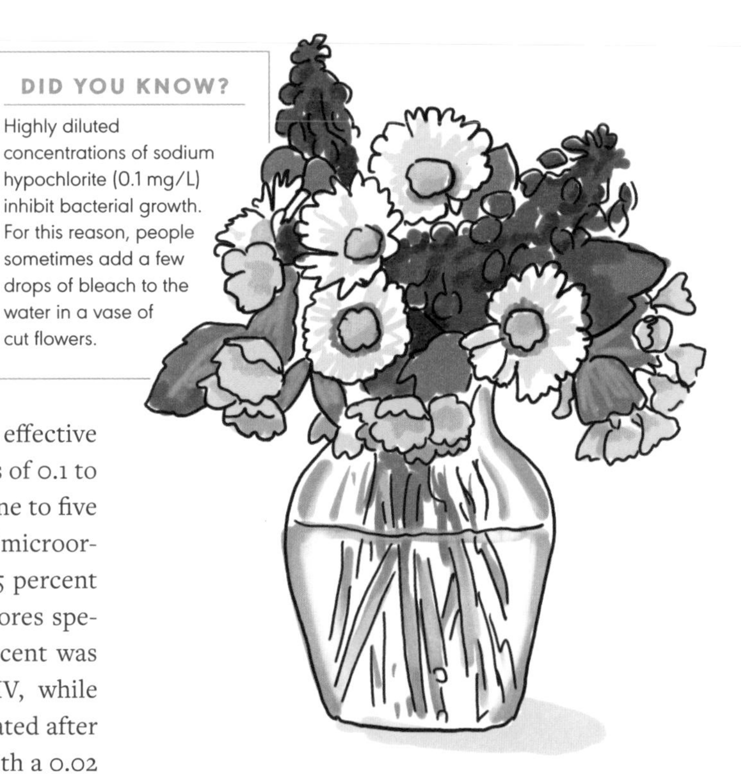

DID YOU KNOW?

Highly diluted concentrations of sodium hypochlorite (0.1 mg/L) inhibit bacterial growth. For this reason, people sometimes add a few drops of bleach to the water in a vase of cut flowers.

Sodium hypochlorite is highly effective against microbes at concentrations of 0.1 to 0.2 percent and contact times of one to five minutes (depending on the type of microorganism), but concentrations of 0.5 percent are recommended for targeting spores specifically. A solution of 0.5 to 1 percent was found to be effective against HIV, while SARS-CoV-2 is completely deactivated after a contact time of thirty seconds with a 0.02 percent solution of sodium hypochlorite obtained by diluting 1 part household bleach in 199 parts water. (Vinegar proved completely ineffective for this purpose.)[10]

Sodium hypochlorite is so useful for getting rid of microbes that various health care institutions worldwide recommend a 0.05-percent solution for disinfecting hard surfaces in non-critical environments (that is, settings that require disinfection but are neither operating rooms nor locations of serious infections like Ebola). If the big bottle of bleach you have at home is a 5-percent solution, you can add 1 part bleach to 99 parts water to create a 0.05-percent solution. In cases where you need to be more thorough—for example, if you need to clean up blood—you can create a 0.5-percent solution by adding 1 part bleach to 9 parts water.

Legally, bleach is considered neither a biocide nor a medical-grade product. As we've seen, that doesn't mean sodium hypochlorite can't kill microbes—bleach is completely indifferent to its lack of registration with the authorities. The reason its effectiveness can't be certified is that, as I explained in chapter 6, the product may degrade over time (so check the expiration date on your household bleach). Also, remember to always rinse with plenty of water after using it, because both the product itself and the sodium chloride it degrades into are potentially corrosive.

10 Catarina F. Almeida et al. "The Efficacy of Common Household Cleaning Agents for SARS-CoV-2 Infection Control." *Viruses* 14, no. 4 (March 2022): 715.

Vinegar

Humans have been wielding vinegar against bacteria for a very long time. The ancient Romans used to apply it to abdominal wounds, and it's been used in food preservation for millennia, so it must have some ability to counter bacterial growth, right? It does: Acetic acid—typically in 5- to 7-percent concentrations—penetrates the cell walls of some microorganisms, causing damage and eventually killing them. This is not exclusively due to acetic acid lowering the pH of the environment, since different acids with equal pH values behave differently.

However, as I've already explained, showing that a substance has bactericidal effects is not enough for it to be considered a disinfectant—the substance must, at the very least, act against a broad spectrum of microorganisms at a given concentration and contact time. Organic acids tend to have less effect against molds and fungi than against bacteria. But since vinegar is a widely used, low-cost product, its potential as a "green" alternative to more classic disinfectants (particularly in parts of the globe where these are not readily available) has led to numerous studies on its effectiveness.

Various websites recommend diluting vinegar in water and using this mixture to clean household objects and surfaces. However, a 1:25 solution failed when tested on two bacteria typically used to validate disinfectants: *Salmonella enterica* and *Staphylococcus aureus*.[11] We shouldn't be surprised by these results, since if such a diluted vinegar was enough to preserve food, it wouldn't be used in its pure or near-pure form to make pickles. But as a chemist, I've noticed the public's unconscious tendency to think about a molecule's properties without considering its concentration, as if these properties were independent of the number of molecules present in a solution. We most often see this cognitive distortion at work when people negatively judge products containing small amounts of substances that cause harm in much greater quantities. We experience the same cognitive glitch when we have positive associations with a substance—this instinctive reaction is often exploited by companies that list ingredients with good reputations (like vinegar) on various packaging even if they appear in such low concentrations that they don't actually do anything.

But, you ask, can I disinfect my kitchen counter with pure, undiluted vinegar? This option has also been tested by comparing it to diluted bleach when used to disinfect ceramic and Formica tiles covered in the kinds of dirt that are often present in kitchens and bathrooms.[12] Pure vinegar was found to be active against the gram-negative bacteria *Salmonella typhi* and *E. coli*, but it proved completely ineffective against gram-positive *Staphylococcus aureus* (as did ammonia), even with a ten-minute contact time. Meanwhile, sodium hypochlorite was effective against all

---

11 Janice M. Bauer, Carol A. Beronio, and Joseph R. Rubino. "Antibacterial activity of environmentally 'green' alternative products tested in standard antimicrobial tests and a simulated in-use assay." *Journal of Environmental Health* 57, no. 7 (March 1995): 13-8.

12 Carole A. Parnes. "Efficacy of sodium hypochlorite bleach and 'alternative' products in preventing transfer of bacteria to and from inanimate surfaces." *Journal of Environmental Health* 59, no. 6 (January-February 1997): 14.

three bacteria in five minutes. If you search for scientific papers investigating this question, you'll find studies that show pure vinegar working against other types of bacteria, such as *Salmonella enterica*,[13] and others testing higher concentrations of acetic acid—for example, by augmenting vinegar with 1.5 percent citric acid.[14] It's notable that vinegar is most often effective against microbes in suspension tests, which involve putting bacteria in solution, while it's drastically inferior against bacteria on the surfaces of fruits and vegetables.

As far as fungi are concerned, vinegar doesn't work—for example, this has been shown in trials on the *Penicillium* mold.[15] Vinegar has also proved ineffective against SARS-CoV-2.[16]

If you don't know the identity of the bacteria in the area you're disinfecting, it's more prudent to use an authorized disinfectant rather than resort to solutions with dubious efficacy. Additionally, the results show that unless you're using an effective disinfectant, the sponge you're applying the product with gets contaminated and transfers bacteria to all the other surfaces it goes on to touch.

None of this means that vinegar isn't useful in certain conditions. For example, it's effective, even when diluted, against the bacterium *Pseudomonas aeruginosa* when disinfecting some types of wounds, and it may prove very useful in treating burn injuries in countries where health care resources are limited.[17] Overall, it's all about understanding the type of microorganism you're up against. It's easy to come up with a list of microbes that vinegar can destroy, and studies on these microorganisms are often quoted by websites about alternative cleaning products. But now you know that if you really want to clean your countertop with vinegar, you'll potentially eliminate some bacteria—you'll just never know how many. And there will be collateral damage, since your kitchen will stink to high heaven for quite some time afterward.

---

13 Daniel Fong et al. *Effectiveness of Alternative Antimicrobial Agents for Disinfection of Hard Surfaces* (Vancouver: National Collaborating Centre for Environmental Health, 2011).

14 Marc-Kevin Zinn and Dirk Bockmühl. "Did granny know best? Evaluating the antibacterial, antifungal and antiviral efficacy of acetic acid for home care procedures." *BMC Microbiology 20*, no. 1 (August 2020): 1–9.

15 John W. Martyny et al. "Aerosolized sodium hypochlorite inhibits viability and allergenicity of mold on building materials." *Journal of Allergy and Clinical Immunology* 116, no. 3 (September 2005): 630–5.

16 Catarina F. Almeida et al. "The Efficacy of Common Household Cleaning Agents for SARS-CoV-2 Infection Control." *Viruses* 14, no. 4 (March 2022): 715.

17 A. P. Fraise et al. "The antibacterial activity and stability of acetic acid." *Journal of Hospital Infection* 84, no. 4 (August 2013): 329–31.

THE ECOLOGICAL COSTS OF DISINFECTION

The first case of the contagious respiratory illness that we would come to know as COVID-19 was confirmed in Wuhan, China, in December 2019. In the months following the WHO's declaration of a pandemic on March 11, 2020, the world witnessed scenes of mass disinfection in indoor and outdoor public places. We saw nebulizer cannons spraying chlorine-based disinfectants like sodium hypochlorite, calcium hypochlorite, chloramine, and chlorine dioxide.

If I close my eyes, I can still see an Italian news report about a town mayor who, perhaps urged on by his frightened constituents, decided to disinfect . . . a beach. Coronaviruses are particularly vulnerable to chlorine-based disinfectants—that's why these substances were recommended from the start, along with appropriately diluted ethyl alcohol, for hospitals and other locations that presented a real danger of transmission. The disinfection furor during that time, fuelled by fear and poor understanding of the mechanisms of contagion, only abated many months later once scientific research proved that the madness was pointless. As I write this book, the obsessive disinfection is still going on in some parts of the world.

When we discussed bleach's whitening properties, I said that judicious household use in the recommended doses aren't a problem for humans or the environment. Well, the practices described in the previous paragraph were anything but judicious: They impacted, and may continue to impact, the environment as well as the health of the many people who were directly exposed, particularly those who worked on the disinfection projects. As I previously pointed out on pages 97 and 104, chlorine-based disinfectants react with organic matter extremely quickly, which is one of the reasons they're so effective. However, when these reactions run wild, it can lead to the formation of toxic and even carcinogenic substances like trihalomethanes, haloacetic acids, acetonitriles, haloketones, and trihalophenols. Only a small fraction of chorine-based disinfectants turn into these substances, but if huge amounts are used, the substances I mentioned can be released into the environment in significant and harmful quantities.

We're only just beginning to see studies on the effects of these treatments, which were carried out all over the world. A 2022 review of all the studies on this phenomenon so far reported that in the months immediately following the first mass disinfections, the concentration of chlorine compounds in Wuhan area lakes rose by up to 0.4 mg/L, which is high enough to negatively impact many aquatic organisms.[18] Once again, we're reminded that harm doesn't necessarily come from a substance's intrinsic properties, but from the way that substance is used.

18 Naseeba Parveen, Shamik Chowdhury, and Sudha Goel. "Environmental impacts of the widespread use of chlorine-based disinfectants during the COVID-19 pandemic." *Environmental Science and Pollution Research* 29, no. 57 (December 2022): 85742–60.

VIRUSES

Not all disinfectants work against viruses. Virucidal activity depends on several factors, some related to a disinfectant's features and others to a virus's structure. All viruses contain genetic material that they use to replicate once they've taken over a host cell, and over the years, we've learned that there are DNA and RNA viruses. These microbes' genetic material is encased inside a protein structure known as a "capsid," although some viruses also have a phospholipid layer around the capsid known as the "pericapsid" or "viral envelope." The latter varieties are called "coated" viruses, while those lacking a viral envelope are "uncoated." The now-iconic images of SARS-CoV-2 show its viral envelope, with the famous protein spikes that allow the virus to dock on human cells and inject them with its own genetic material.

Coated viruses tend to be more vulnerable to disinfectants than uncoated viruses. Ethyl alcohol in particular is very effective against coronaviruses and other coated viruses at a concentration of 70 percent, while it's completely powerless against uncoated viruses such as norovirus, the most common cause of acute non-bacterial gastroenteritis.[19]

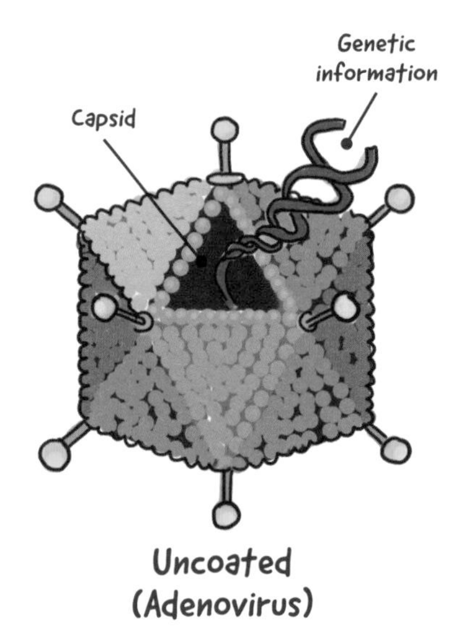

Uncoated (Adenovirus)

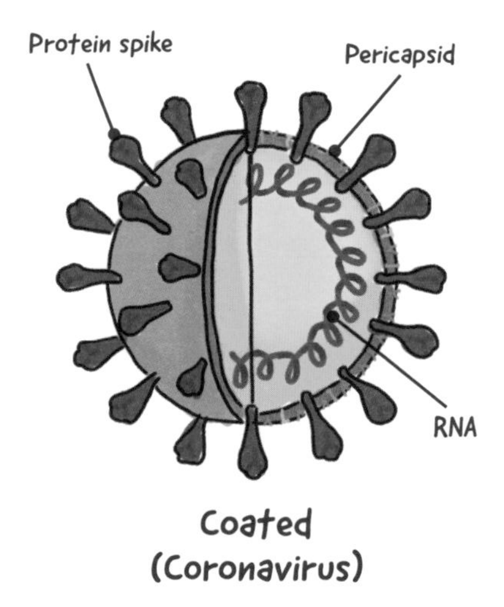

Coated (Coronavirus)

19 Every now and then, the news spotlight falls on norovirus when the kitchen of a well-known restaurant is found to be contaminated, forcing it to close for a few days. This group of viruses is responsible for some 685 million infections worldwide every year. Of these, about 200 million affect children under five, causing 50,000 deaths in developing countries. "Norovirus Burden and Trends," Centers for Disease Control and Prevention, last modified May 8, 2023, cdc.gov/norovirus/burden.html.

THE MOTHER OF ALL KITCHEN CONTAMINATION

I'd like you to walk into your kitchen and look at all the objects it contains. Which one do you think is the most contaminated with bacteria—the one that's most likely to cause a foodborne infection? I'm sure many of you will have guessed: It's the humble kitchen sponge. (The dishcloth is also a contender.) In the US alone, the CDC estimates that foodborne bacterial contamination causes 48 million cases of disease every year.[20] We may become infected at restaurants or at school and work cafeterias, but most of these cases originate in our homes. Some studies estimate that 87 percent of these infections come from household kitchens,[21] and the number one suspect is none other than the sponge we use to clean dishes and cooking surfaces.

Confess: How often do you disinfect your sponges? Never, perhaps? Truly impressive microbial menageries have been found on domestic kitchen sponges: *Salmonella*, *Campylobacter*, *Staphylococcus*, *Escherichia*, and *Listeria*—all bacteria responsible for potentially serious infections and diseases (and I'm sparing you from a list of the viruses). One study reported that *Salmonella* was present on 15 percent of American kitchen sponges and 14 percent of dishcloths.[22] From sponges and dishcloths, bacteria can be transferred to plates, glasses, and utensils, and then to us. But isn't there anything we can do? Well, of course there is: Ditch your sponges without mercy once they're worn out, and until then, disinfect them regularly. Let's see how.

In 2009, three FDA researchers published a paper on the best way to disinfect kitchen sponges.[23] First, the researchers saturated the sponges with beef broth to create a good growth medium and then inoculated them with bacteria before leaving them to incubate for two days. Then, the sponges were subjected to various cleaning methods: soaking them in a 10-percent bleach solution for three minutes or in lemon juice for one minute, sending them through a dishwasher cycle, or putting them in the microwave at full power for one minute. After that, the sponges were examined for any residual contamination with fungi or bacteria.

And the winner was . . . the microwave, which turned out to be even more effective against bacteria than bleach, while the

20 "Burden of Foodborne Illness: Overview," Centers for Disease Control and Prevention, last modified November 5, 2018, cdc.gov/foodborneburden/estimates-overview.html.

21 Elizabeth C. Redmond and Christopher J. Griffith. "Consumer food handling in the home: a review of food safety studies." *Journal of Food Protection* 66, no. 1 (January 2003): 130–61.

22 C. Enriquez et al. "Bacteriological survey of used cellulose sponges and cotton dishcloths from domestic kitchens." *Dairy, Food and Environmental Sanitation* 17 (1997): 20–4.

23 Manan Sharma, Janet Eastridge, and Cheryl Mudd. "Effective household disinfection methods of kitchen sponges." *Food Control* 20, no. 3 (March 2009): 310–3.

second best treatment was the dishwasher. The latter was in fact slightly better against fungi than the microwave, with both methods working much more effectively than bleach.

While this experiment used a 1300-watt microwave, one minute at the highest setting in less powerful appliances is enough to deactivate most, if not all, microorganisms. But there are two things to bear in mind before you put your kitchen sponge in the microwave. First, you must be sure that the sponge doesn't contain any metal, since you really don't want to risk setting off an electric discharge. Second, you should get the sponge completely wet, making sure that it won't dry out before the minute is up, and set it in a microwave-safe container rather than directly on the turntable. (And while we're on the subject, remember to clean your microwave often!)

The microwave's disinfecting action probably comes from to two factors: the high temperature of 176°F (80°C) reached by the water inside the sponge, and radiation's ability to kill bacteria by damaging their DNA.

You might be surprised that soaking a sponge in a 10-percent bleach solution (with an actual bleach concentration of 0.525 percent) is only slightly more effective than soaking it in water alone. Aren't similar concentrations of sodium hypochlorite recommended for disinfecting surfaces, even in health care settings? This is a beautiful example of theory and practice not always walking hand in hand. (In Yogi Berra's immortal words, "In theory, there is no difference between practice and theory. In practice, there is.") Here we see that with a consistent microbial load, a disinfectant's action is heavily dependent on the material it's disinfecting. The researchers speculated that the bleach solution may have failed due to a large amount of organic material in the sponge, which, you'll remember, was soaked in beef broth. Since sodium hypochlorite is highly

WARNING!

To recap: NEVER put a dry sponge in the microwave. The appliance requires water to absorb its microwaves, which can otherwise cause permanent damage. If you want to use this disinfection method, get the sponge soaking wet, place it in a container, and be very careful not to burn yourself when you remove it.

reactive with organic substances, only a small part of it was able to reach the microbes, while the rest oxidated the broth's organic molecules.

You may be thinking that your kitchen sponges, filthy and fetid as they may be, are not exactly impregnated with beef broth—after all, you do rinse them under the faucet and give them a good squeeze before setting them down. Well, that's fair. An earlier study from 1999 involved researchers collecting real kitchen sponges that volunteers used at home for one or two weeks.[24] In this case, too, the sponges were inoculated with bacteria and left to incubate before being subjected to various treatments. Once again, putting the sponge in the microwave at full power led to a 99.999 percent reduction in microbes, and a similar figure was found with the dishwasher cycle. However, first place went to boiling the sponge in water, while treating it with a 5-percent bleach solution was as effective as the microwave or dishwasher. It seems that in this case, the hypochlorite could work undisturbed. The treatments that didn't work included a washing machine cycle with regular laundry (unless bleach was added), hydrogen peroxide, ammonia, and vinegar.

Since I mentioned dishcloths at the beginning of this section, you'll be happy to hear that they, too, have been the subject of an experiment. Let's jump straight to the results: Microwaving wet dishcloths at full power (in this case, 700 watts) once again completely eliminates the microbial load.[25] The only problem here is size: The dishcloths used in this trial measured about 2.5 by 4.75 inches (or 6 by 12 cm), and I can't see myself stuffing one of my much larger dishcloths into a glass to place it in the microwave. I think I'll carry on changing them frequently and putting the dirty ones through the washing machine at a minimum temperature of 140°F (or 60°C) followed by the dryer.

DID YOU KNOW?

Along with disinfecting your kitchen sponge regularly, it's also best not to always leave it wet. Humidity, a warm environment, and the presence of organic residue combine to make an excellent microbial culture medium. A little trick you may find useful is having at least two sponges and swapping them every other day so that one has time to properly dry out while you're using the other. This practice is no substitute for regular disinfection, but it at least slows down the multiplication of mold and bacteria.

24 Judy Y. Ikawa and Jonathan S. Rossen. "Reducing bacteria in household sponges." *Journal of Environmental Health* 62, no. 1 (July-August 1999): 18

25 Emily Gillies. "Determining the most effective common household disinfection method to reduce the microbial load on domestic dishcloths: a pilot study." *Environmental Health Review* 63, no. 4 (December 2020): 101-6.

TYPES OF MICROORGANISMS[26]

DISINFECTANTS

	GRAM+ BACTERIA	GRAM- BACTERIA	GRAM+ MICRO-BACTERIA	FUNGI	VIRUSES	BACTERIAL SPORES
Sodium hypochlorite	Effective	Effective	Effective	Effective	Effective	Effective
Phenolic compounds*	Effective	Effective	Partially effective	Effective	Partially effective	Not effective
Quaternary ammonium compounds (QACs)	Effective	Partially effective	Not effective	Effective	Partially effective	Not effective
Formaldehyde or formalin*	Effective	Effective	Partially effective	Effective	Effective	Partially effective
Glutaraldehyde*	Effective	Effective	Effective	Effective	Effective	Effective
Hydrogen peroxide	Effective	Effective	Effective at high concentrations	Partially effective	Partially effective	Effective at high concentrations
Iodine compounds*	Effective	Effective	Effective	Effective	Effective	Effective
Alcohols	Effective	Effective	Partially effective	Partially effective	Partially effective	Not effective
Chloramine*	Effective	Effective	Effective at high concentrations	Effective at high concentrations	Effective at high concentrations	Effective at high concentrations
Chlorhexidine*	Effective	Partially effective	Not effective	Effective		Not effective
Peracetic acid*	Effective	Effective	Effective	Effective	Effective	Effective

KEY

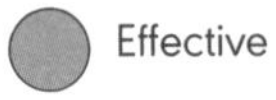

Effective

Partially effective

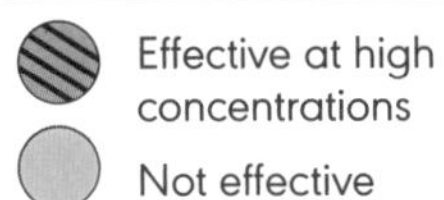

Effective at high concentrations

Not effective

*Not suitable for household use

26 Adapted from the Italian publication "La sanificazione nell'industria alimentare e negli allevamenti," produced by the Modena Local Health Authority, Regional Health Service of Emilia-Romagna.

MOLD

I absolutely love mushrooms, as long as they're the right species. I find *Cantharellus cibarius*[27] to be best in risotto, even though I'm not against using *Boletus edulis*.[28] As for *Macrolepiota procera*, it's just begging to be breaded and fried. And what about non-mushroom fungi like *Penicillium roqueforti*, without which we wouldn't have blue cheeses like gorgonzola?

But far be it from me to hold a purely gastronomic (and anthropocentric) view of fungi, because this kingdom encompasses a vast number of species that play crucial roles in ecosystems. In the woods, for example, fungi decompose fallen leaves and dead trees, recycling organic matter—when you're on a walk and see a log covered in mold, you're looking at a fungus.

Molds are fungi, and there are many, many different kinds—or, more accurately, genera: *Aspergillus*, *Cladosporium*, *Alternaria*, and *Penicillium*, for instance. As long as they stay outdoors, we're all good. If, however, we come into contact with them—apart from the few varieties that produce delicious food—things won't go well.

We all know that mold growing on food where it doesn't belong can be very dangerous. Some molds produce substances known as microtoxins, which can cause very serious problems if they're ingested, including immunodeficiency, neurological damage, and kidney and liver cancer. These toxins may be present on grains (like corn), dried or dehydrated fruits (like pistachios or figs), and many other foods. Unfortunately, the fungi responsible aren't always visible to the naked eye, so sometimes we just need to trust the regulations that exist to keep our food safe.

Besides the molds that can be ingested, there are also varieties that can be inhaled (or rather, their spores can). Mold spores are virtually everywhere, and it's simply

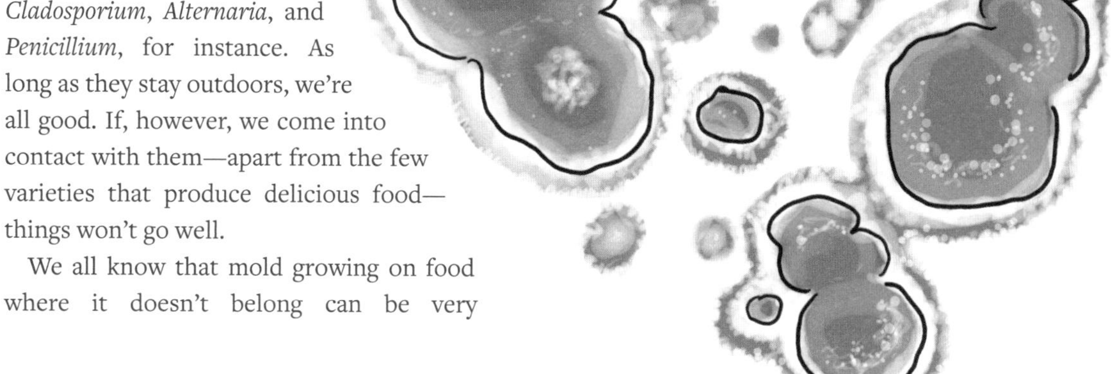

27 Known by dozens of common names, including "chanterelle" and "girolle."

28 Commonly known as "porcini."

impossible to keep them all out of our homes. They don't tend to be a problem until they end up in an environment that's constantly damp or wet—that's when mold begins to colonize the area, sometimes growing very fast indeed.

Basements, bathrooms, showers, the undersides of sinks, the rubber seals of fridges and washing machines, and any other places that harbor persistent humidity are all danger zones. The most effective way to combat household mold colonies is to prevent humidity in these areas by ensuring good ventilation and keeping damp spots as dry as possible. If you have a broken roof tile that's causing water leakage and you notice a dark stain in the corner of your bedroom ceiling, or if a pipe is dripping and causing a greenish growth to blossom on the wall, you'll want to tackle the source of the damp, because otherwise, you can depend on the mold coming back again and again.

Living in an environment that's been permanently colonized by mold has been linked to innumerable health problems, mostly related to the respiratory system: allergies, asthma, rhinitis, and conjunctivitis, among others. Also, children who grow up in moldy environments are more likely to develop asthma. So, if there's mold in your living space, don't underestimate it—deal with it as soon as you can.

If you've ever walked into a house that's been neglected for a long time, you'll notice a moldy odor that hits you as soon as you open the door. Once you open the windows, you can see the disaster unfolding on the ceiling, walls, and perhaps even the couch. Everything needs to be sanitized. But aren't fungi microorganisms? A disinfectant can kill them, right? Well, that's the theory, but it's not so simple.

If you find yourself in this regrettable situation, the first thing on your list should be assessing whether you can do the cleaning job yourself or if you need to call in a professional. Because of the spores that can get knocked loose when cleaning and enter your lungs, I would never work on a large moldy area without a face mask, preferably an FFP3. And if the area is larger than one square meter, it's probably best to find someone who deals with this kind of thing for a living.[29] They won't

WARNING! Bleach kills microorganisms very quickly, and that includes mold. If the bleach you bought for your laundry isn't concentrated enough, there are products on the market containing chlorine compounds (like sodium hypochlorite or sodium chlorate) in higher concentrations, sometimes over 10 percent. Be very careful when handling these products: Always remember to wear gloves, a face mask, and safety goggles. If the moldy area is very large, or if you end up back at square one a few months after extensive treatment, you'll probably need to eliminate the source of the humidity and treat the walls with special products—and for that, I recommend hiring a professional.

29 According to the EPA's recommendations at epa.gov/mold.

just remove the mold, but also eliminate the root cause of the damp that allowed it grow.

For smaller areas, like one corner of a room or one drawer of a dresser, you can do the work yourself using one of the many products on the market, all of which have one or more disinfectants as their active ingredients. The most common is probably sodium hypochlorite, which is also the active ingredient in household bleach and many other disinfectants. For a long time, however, there were conflicting opinions about its use. Let's find out why.

We know beyond a doubt that hypochlorite works against molds. A 2012 study showed that a commercial spray containing sodium hypochlorite at a 2.4-percent concentration (slightly less concentrated than the household bleach sold in stores) can eradicate the molds of various fungi varieties such as *Aspergillus*, *Cladosporium*, and *Alternaria*.[30] This spray worked on both glazed and porous ceramic tiles after a contact time of five minutes.[31]

There are two problems with this finding. The first is that respiratory and immune problems caused by mold can occur even if the fungus is dead, because allergens can remain intact on spores' surfaces. This is one of the reasons why in the past (and even now, to an extent), some authorities advised against using hypochlorite: The risks were not worth the advantages. The concern was that hypochlorite and similar products could simply bleach mold without eliminating its allergenic potential.[32]

Fortunately, this doesn't seem to be true. A hypochlorite solution sprayed on construction material colonized by *Aspergillus fumigatus* not only killed the fungus itself, but also eliminated most of the spores' allergenic potential.[33] The result was verified both in the lab (using the ELISA test, a classic biochemistry assay) and by volunteers before getting confirmed in the 2012 study I mentioned above, which covered a broader sample of fungi. In conclusion: not only does hypochlorite kill fungi, but it also deactivates their spores.

The second problem is that the fungus may have penetrated the surface of the tiles to such an extent that a disinfectant can never reach the whole colony. In a 2013 study, pinewood blocks were colonized with fungi typically found in our homes by incubating them in a humid environment at 77°F (25°C) for eight weeks.[34] They were then treated with various disinfectants, including hydrogen peroxide at 17-percent concentration, hypochlorite at 6.15-percent concentration, quaternary ammonium salts, and phenols. All these products were effective at killing

---

30 Kelly A. Reynolds et al. "Occurrence of household mold and efficacy of sodium hypochlorite disinfectant." *Journal of Occupational and Environmental Hygiene* 9, no. 11 (2012): 663-9.

31 More precisely, the reduction observed was greater than 6-log.

32 Lynnette J. Mazur and Janice Kim. "Spectrum of noninfectious health effects from molds." *Pediatrics* 118, no. 6 (December 2006) e1909-26.

33 John W. Martyny et al. "Aerosolized sodium hypochlorite inhibits viability and allergenicity of mold on building materials." *Journal of Allergy and Clinical Immunology* 116, no. 3 (September 2005): 630-5.

34 P. Chakravarty and Brad Kovar. "Engineering Case Report: Evaluation of Five Antifungal Agents Used in Remediation Practices Against Six Common Indoor Fungal Species." *Journal of Occupational and Environmental Hygiene* 10, no. 1 (2013) D11-6.

the mold. Then, the wooden blocks were left to dry out for five weeks before samples were harvested from the surface and placed in a culture medium. In every case, the fungus had begun to grow back. It seems that the disinfectants were unable to kill it all, and the remaining spores were reactivated as soon as they ended up in a suitable environment.

This is unfortunately often the case with porous materials colonized by mold, whether they're wood, grout, or plaster, and the only solution may be demolition. For this reason, even if you can probably clean shower tiles or perhaps even a small part of a wall, don't wear yourself out trying to remove mold from wooden frames or doors.

USING DISINFECTANTS SAFELY[35]

35 Adapted from the Italian publication "I biocidi: Quaderni per la salute e la Sicurezza," produced by the National Institute for Insurance against Accidents at Work (INAIL).

11

Baking Soda

Using a product—no matter how harmless it is—for no good reason is no different from throwing it away and polluting the environment unnecessarily. It also wastes the many resources that went into creating the product in the first place, from the energy and raw materials used in its chemical synthesis to the packaging and shipping required to get it into our hands. That's why I'm so annoyed when I see people recommending baking soda for things it absolutely can't help with.

Therefore, I've decided to bring together all the common baking soda tips and tricks (both right and wrong) in one place. If I've gone into more detail about something in previous chapters, I summarize it here.

But before I begin, I'd like to ask you to stop and consider something. Don't you think it's a bit weird how these miraculous natural remedies touted throughout the internet and plastered across household magazines all seem to focus on what I call "the four horsemen of the cleaning apocalypse": vinegar, salt, lemon, and baking soda? If these four ingredients were really all we needed, the chemists who focus their life's work on developing cleaning products would probably have dedicated themselves to something else.

And why is baking soda in particular so popular? One reason is the theatrical but completely functionless fizziness it produces when mixed with vinegar. All those bubbles suggest that something very important is occurring. And yes, something is most definitely happening: Carbon dioxide is being released. But that's all—nothing more, nothing less.

The second reason is a little more nuanced: Humans don't like uncertainty, and we can't stand not having a fix for a problem or not looking for answers when the solution isn't obvious. Using an inexpensive, easily accessible, ostensibly harmless ingredient to get a stain out of our favorite tablecloth, render our cat's litterbox odor-free, and "disinfect" our mattress makes us feel like we've taken control of finding a solution. It's much more satisfying than using a specific chemical product with a composition we don't understand, even when we read the list of ingredients.

People tell me, "Baking soda worked for me!" No, it didn't work—it really didn't. If you truly want to know how effective a product actually is—objectively, unbiased by external expectations—it's crucial to compare it with another cleaner under exactly the same circumstances. What's more, to avoid any unconscious bias, the test has to be a "blind trial": We should be completely unaware of which item received the baking soda treatment and which one didn't. The human brain is very good at convincing us that, yes, all things considered, the tablecloth looks much better after being cleaned with baking soda, especially if we want this to be true and have nothing to compare the cleanliness with.

Isn't chemistry great? It helps us make sense of the world and protects us from anyone trying to sell us false promises.

WHAT *CAN* BAKING SODA DO?

Whenever I debunk claims about baking soda's effectiveness as miracle cure, without fail, I get asked, "What *does* it do, then?" Well, let me tell you, the list of things baking soda can do is pretty impressive.

Neutralizing acids

Let's start with the most obvious characteristic: Sodium bicarbonate is a weak alkaline substance that reacts with and neutralizes acids. This is why, as I've explained before, vinegar and baking soda essentially cancel each other out when they're mixed. So, if you need to deactivate an acid, using baking soda either in its powder form or dissolved in water is a good move. For example, baking soda can work wonders if some rotten tomatoes are oozing gunk in the fridge, creating a playground for bacteria that produce even more foul-smelling acidic liquids.

Our stomach is also an acidic environment, and sometimes it develops an excess of hydrochloric acid. Many antacids have sodium bicarbonate as one of their main ingredients. When you swallow one of these, the bicarbonate yields carbon dioxide, provoking a big, therapeutic burp!

Producing carbon dioxide

The carbon dioxide produced when baking soda reacts with an acid can be used in a variety of ways. For example, before sparkling water made its way onto supermarket shelves, some people used to dissolve sachets of powder containing baking soda along with one or more solid acids (like cream of tartar, citric acid, or malic acid) in tap water to get the same fizzy effect. Spooning the powder into a bottle of water and screwing the cap on instantly caused carbon dioxide to bubble, then dissolve in the water since it had nowhere else to go. This drink had a bit of an aftertaste, though, as the bicarbonate also yields salt when it reacts with acid.

This reaction, often between the same substances, is how instant or "chemical" yeasts help cakes and cookies to rise. These acidic substances have to be dosed accurately to avoid any nasty surprises when you take the cake out of the oven, since the wrong ratios can leave you with a much darker product than you expected or a soapy taste when you take a bite.

If a recipe specifies a chemical leavening agent, use a chemical leavening agent, since replacing it with just baking soda falls short of the acid required to produce the necessary amount of carbon dioxide bubbles. When a recipe calls for baking soda alone, the

DID YOU KNOW?

Like all salts, baking soda is not volatile. This means that when it's dissolved in hot water, the vapor it produces is made up of water alone, since the bicarbonate itself remains in solution. The same thing happens with table salt: The steam rising from the pot when cooking pasta is not salty—unless a bit of the actual cooking water, which does contain salt, bubbles over.

ingredients likely include sufficient quantities of acidic foods like honey, yogurt, and lemon juice to make up the difference.

Baking soda produces carbon dioxide on contact with acid and with water at a temperature higher than 176°F (80°C). Try tipping a teaspoon of sodium bicarbonate into some very hot water, then stand back and watch the CO_2 fizz. This is a simple reaction, but it's not great for cake-baking, as the carbon dioxide generated for leavening power is only half of what can be obtained when baking soda reacts with an acid. Also, the sodium carbonate salt left over from the hot-water reaction is far more alkaline than the original bicarbonate, which can interfere with the other chemical reactions that take place while the cake is cooking.

The reaction we've been discussing is relatively slow until the temperature gets above 212°F (100°C) and decomposition speeds up significantly. This is how candy manufacturers get bubbles into molten sugar. When I was a child, a common Christmas stocking filler was black candy that looked like real coal. As I munched on it, I'd often wonder how they managed to create such delicious, easy-to-crunch air bubbles. The secret is heating the molten sugar to 284°F (140°C), adding baking soda, and stirring vigorously. The bicarbonate instantly decomposes, and when the sugar cools, the bubbles of CO_2 that get released are trapped inside the hardened candy. At 284°F (140°C), molten sugar is still clear, since caramelization has yet to take place. This is why my candy coal tasted like sugar despite its black color, which came from added vegetable charcoal. Honeycomb toffee (or hokey pokey, as it's known in some places) is based on the

same principle, only the baking soda is added at 302°F (150°C), when the sugar takes on that unmistakably rich, enticing smell as it turns into caramel.

Eliminating gross smells

Acidic substances sometimes cause foul odors—one nasty example comes from the fatty acids produced by the bacteria in our armpits or on our feet. Some people prefer rubbing a little baking soda under their armpits instead of using everyday antiperspirant deodorant, but bicarbonate can't eliminate all stinky smells of microbial origin, especially since we all have different bacterial makeups that change over time.

Baking soda can help a bit with smelly feet, too. Try sprinkling some inside any particularly malodorous shoes or slippers, leaving it there for a day, then sucking up the remaining powder with a vacuum cleaner. I can't promise it will be 100 percent effective, but it's a temporary fix.

However, baking soda can't help with unpleasant smells that come from alkaline sources—like the smell of rotting fish, which is caused by compounds called amines. At best, after you've removed the culprit from your fridge, you can sprinkle some bicarbonate on the leftover liquid, although I also recommend wiping the surface down with water plus vinegar or citric acid to remove any amines still lurking in the grooves and crevices of the refrigerator shelves.

Baking soda also absorbs liquids, so a little bit sprinkled on the bottom of your trash can will soak up any sludge, reducing its volatility and getting rid of its odor.

Softening vegetables

Adding a teaspoon of baking soda to the water when boiling legumes and vegetables makes them cook and soften more quickly. It's not so much the bicarbonate that does the job, but the alkalinity of the water. Conversely, in an acidic environment, the same beans would be harder and take longer to cook.

You can use this technique in all kinds of ways. To make the perfect cream of onion soup, try adding a pinch of baking soda to the cooking water—the onions will soften in no time. For the crispiest roast potatoes, boil them with a little baking soda before baking.

Raising pH

Sometimes, you need to raise a solution's pH to get the effect you're looking for. As an example, the Maillard reaction (which makes cooked food browned and tasty) happens more readily at a high pH. This means that adding a pinch of baking soda to carrots and pumpkin in the oven before blending them into a velouté sauce intensifies the flavor.

Also, a higher pH can help vegetables like green beans keep their vibrant color when they're boiled. In an acidic environment, they tend to lose their luster very quickly. Be careful, though, since as we just explored, baking soda also softens the beans much faster. And don't forget that although bicarbonate is a weak base, it turns into the much more alkaline sodium carbonate when it's heated above 176°F (80°C), so go easy on the quantity.

Acting as an abrasive

Baking soda's secret scouring power makes it helpful to have around the house and explains why it's the abrasive of choice for professionally cleaning objects like statues. Table salt can also be used for the same purpose, but with two caveats: It's highly soluble in water, and its grains are much bigger. I use salt for getting rid of the burnt crust from the bottom of my pots, but not much else. Baking soda, on the other hand, is not particularly soluble in water and is so fine that it's practically imperceptible.

Try wetting baking soda with a little water, spreading it on oven trays, tiles, or whatever it is you want to clean, and then scrubbing it with a sponge. Some people recommend moistening it with vinegar, I know, but this wastes both vinegar and baking soda. To capitalize on baking soda's abrasive power, all you need is a little water or, in some cases, another liquid that doesn't react with the bicarbonate. You can choose substances like alcohol, oil, dish soap, or glycerin if the residue you want to remove is susceptible to one of them.

No matter what, always check before you clean to make sure baking soda won't damage the surface. Oven walls, for example, have a thin coating that this substance can scratch off.

WHAT *CAN'T* BAKING SODA DO?

I think I've mentioned baking soda in more than half of the chapters in this book, usually to debunk some nonsense claim. I'll summarize all those points here for easy reference.

Cleaning or removing grease

By now, you know everything about detergents and surfactants. Baking soda is neither, and it can't remove greasy dirt the way soap or detergent can (not even when mixed with vinegar, as internet sources commonly claim). I've seen videos where baking soda is mixed with dish soap, but it's still surfactants doing the cleaning work—baking soda's only role is that of an abrasive.

Bleaching or killing bacteria

Of all the misconceptions surrounding baking soda, this is the one that amazes me the most. I've even heard recommendations for adding it to the washing machine as a bleaching agent. This makes absolutely no sense. A bleaching agent is a generally an oxidizer based on sodium hypochlorite or hydrogen peroxide that can destroy colored molecules, but baking soda has no such ability. If you're feeling doubtful, think about this: We can't drink bleaching agents because they're hazardous substances, but we can taste baking soda without being harmed. It's harmless because unlike bleach, it's not an active agent. And with no bleaching ability, it clearly doesn't have the antibacterial properties of real bleach, either.

Eliminating limescale or softening water

By now, we know that limescale is an alkaline deposit and acid is required to dislodge it. Baking soda is also alkaline, so it can't do anything to limescale. This isn't to say you can't use it as an abrasive for scrubbing your faucets, but you'd get the job done much quicker with a specialized limescale remover.

It's also inaccurate to suggest adding baking soda to your washing machine's drum to prevent limescale formation. Unfortunately, this recommendation has crept into the occasional washing machine manual, making me seriously doubt the reliability of the manufacturer and their products. The only way to prevent limescale is to use something that eliminates calcium ions or captures them and carries them away, but baking soda doesn't have this power.

In this case, I can guess where the misconception might have come from. In times gone by, when people boiled their laundry and didn't have easy access to sodium carbonate, they might've made do with baking soda. It's easy to understand why: I mentioned earlier that, at 176°F (80°C) and above, baking soda converts to sodium carbonate. This substance precipitates calcium carbonate in the presence of calcium ions. That said, no one does laundry at these temperatures anymore, and even if they did, modern detergents already include water softeners.

Find and remove the source of the bad smell

Trapping bad smells

We saw how baking soda reacts with smelly acidic substances to neutralize them, but simply putting a bowl of baking soda in the fridge does not capture lingering smells—it's not an odor extractor. If an acid molecule were to randomly dive headlong into the baking soda, it would probably get trapped, but the likelihood of all the particles responsible for a stench

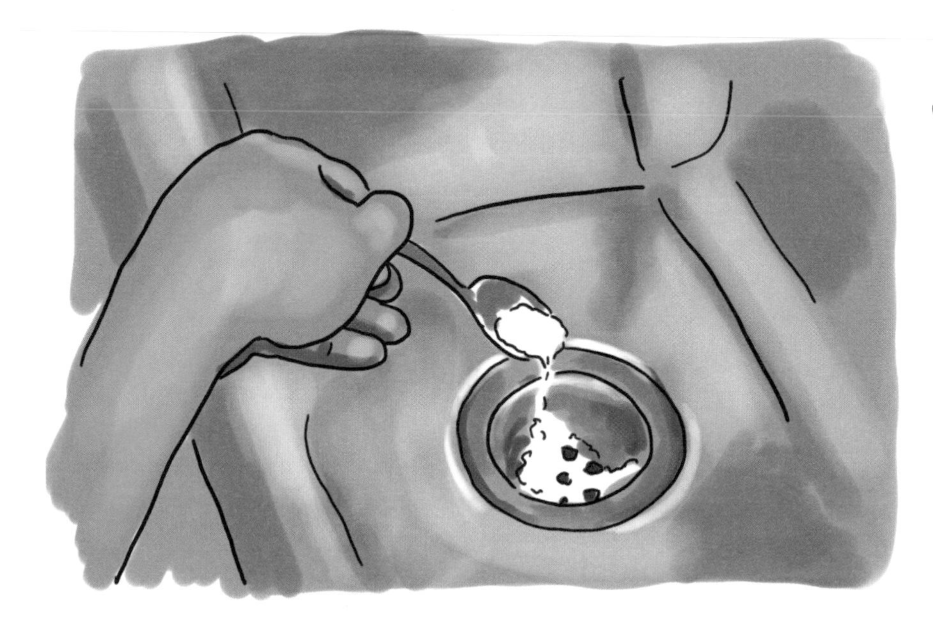

Baking soda will not unclog your sink

ending up in that bowl as they move around the fridge is close to zero. Not to mention the fact that many foul-smelling molecules are not even acidic. My advice is to find the root cause of the stink and deal with it directly.

Unblocking sinks

A hugely popular yet incorrect and potentially dangerous recommendation is to use baking soda to unclog a backed-up sink, in combination with vinegar and often with table salt thrown in for good measure. As I explained earlier, baking soda has no chemical properties that make it effective against the likely sources of the blockage: limescale, fats, and proteins.

Washing hair

Hairs are covered with overlapping cells, like fish scales, that lift up or face downward depending on the pH of their environment. In the presence of an acid, they flatten completely, making the cuticle smooth to the touch, while in an alkaline environment, the cells lift up, causing friction between hairs. When they can no longer slide over each other easily, your hair looks messy and feels rough. Therefore, most modern shampoos have acidic pH values to flatten the cuticle cells. This is also why it's not generally advised to wash your hair with soap, which has a very basic pH.

Disinfecting

Baking soda is not a disinfectant. I know countless people, some of whom are even doctors, who claim otherwise—but they're wrong. Some baking soda brands even state on their packaging that baking soda has no disinfectant properties. Nevertheless, this hoax has taken a firm hold thanks to word of mouth and the mind-boggling number of people who spread it around the internet.

This doesn't mean that there aren't bacteria somewhere that can be eliminated by baking soda under the right conditions.[1] (In fact, this is probably true for almost every chemical substance.) But to qualify as a disinfectant, a chemical agent must be effective against a wide range of microorganisms, especially those responsible for illnesses.

If you need further proof after reading about disinfectants in chapter 10, consider this study. A group of scientists assessed the disinfectant capability of a number of household products—including bleach, hydrogen peroxide, vinegar, and baking soda—against three of the most common bacteria that cause food poisoning: *Listeria monocytogenes, Escherichia coli* O157:H7, and *Salmonella typhimurium*.[2] They carried out suspension tests by adding the products directly to the microorganisms, which were suspended in liquid in test tubes. As I've mentioned before, there's no guarantee that a product that's effective at certain concentrations and exposure times against bacteria in suspension will be equally effective against the same bacteria deposited on a head of broccoli or an apple. But if the substance fails even in suspension, then you can be sure it's useless.

Well, that's exactly what the scientists found when they examined solutions of baking soda at increasingly higher concentrations. Even at 50 percent (240 grams of baking soda and 240 grams of water), the solution was completely ineffective and showed no antimicrobial activity. The researchers were not surprised by this—it merely confirmed the findings of previous studies. Unfortunately, the hoax rages on all the same.

In the same study, the researchers found that the weakest 3-percent sodium hypochlorite solution was effective at destroying all three types of bacteria at room temperature in less than sixty seconds. In second place was 3-percent hydrogen peroxide (the most common commercially available concentration). The solution had to be heated to 131°F (55°C) to eliminate *Listeria*. Undiluted vinegar (a 5-percent acetic acid solution) was similarly effective against *Salmonella* at room temperature in less than sixty seconds, but it also had to be heated to 131°F (55°C) to kill *E. coli* and *Listeria*. A 5-percent citric acid solution was less useful because it had to be heated in all three cases and applied for ten minutes to do any damage to *E. coli* and *Listeria*.

To sum it all up, the researchers emphasized that the results were obtained from bacteria in suspension, and microorganisms are easier to render inactive in this situation. Therefore, their conclusions do not support products' effectiveness on actual surfaces and food in a real home environment.

1 For example, one study showed that baking soda eliminates *Feline calicivirus* from a steel surface: Yashpal S. Malik and Sagar M. Goyal. "Virucidal efficacy of sodium bicarbonate on a food contact surface against feline calicivirus, a norovirus surrogate." *International Journal of Food Microbiology* 109, no. 1-2 (May 2006): 160-3.

2 Hua Yang et al. "Inactivation of Listeria monocytogenes, Escherichia coli O157: H7, and Salmonella typhimurium with compounds available in households." *Journal of Food Protection* 72, no. 6 (June 2009): 1201-8.

WASHING FRUITS AND VEGETABLES

If baking soda isn't helpful for making sure your produce is clean, what should you use instead? The quick and easy answer is to rinse fruits and veggies under the faucet. If you leave them to soak, you could actually end up spreading contamination—for example, a questionable bit of lettuce could spoil all the other healthy leaves. If you'd still rather soak your produce, then remember to rinse it afterward. And if your doctor has recommended using a sodium hypochlorite solution (whether due to pregnancy or some other health reason), then go ahead. Otherwise, washing with water is more than good enough.

Three scientists from the University of Georgia Food Science & Technology department tested the efficacy of various consumer-friendly washing products and technologies for reducing pathogens on fresh produce.[3] They inoculated tomatoes, broccoli, melons, and lettuce with three bacterial strains commonly associated with food poisoning: *Salmonella enterica*, *Escherichia coli* O157:H7, and *Listeria monocytogenes*. Then, they cleaned the fruits and vegetables using different substances: ozonated water, diluted chlorine bleach solution, electrolyzed oxidizing water,[4] and commercial produce wash, all of which were compared with ordinary tap water. After this, the researchers measured the log reduction in bacterial contamination.

All of the treatments effectively lowered surface bacterial contamination, although none of them eliminated enough microbes to be truly considered disinfectants.[5] On average, the chlorine bleach solution was the most effective, but it didn't work equally well on all vegetable–pathogen combinations. Smooth-skinned produce like tomatoes and green onions were easier to clean than broccoli (which has a more uneven surface structure) or spinach (which has a surface film that repels water). Likewise, some washing treatments were more useful than others for certain vegetables. On tomatoes, for example, chlorine bleach water effectively eliminated both *E. coli* and *Listeria*, while running tap water was more effective than ozonated water and electrolyzed oxidizing water at killing *E. coli* and just as good at rendering *Salmonella* and *Listeria* inactive. Broccoli and melon had similar results.

The researchers concluded that their results "suggest that each of the washing treatments tested has the potential to reduce surface bacterial contamination on specific items of fresh produce, but none produced significantly greater reductions than tap water rinse for all tested items. . . . For some

3 Jillian D. Fishburn, Yanjie Tang, and Joseph F. Frank. "Efficacy of various consumer-friendly produce washing technologies in reducing pathogens on fresh produce." *Food Protection Trends* 32, no. 8 (August 2012): 456-66

4 The devices that produce this substance generate a solution of sodium hypochlorite and sodium hydroxide.

5 As we've learned, disinfection generally requires a reduction in bacterial load of 4-log or 99.99 percent.

produce/pathogen combinations, running tap water was as effective as the commercial technologies."

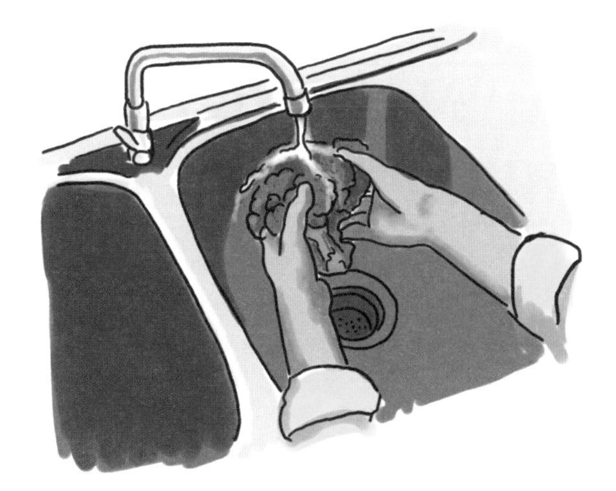

Another study used *Listeria innocua* to contaminate tomatoes, apples, lettuce, and broccoli, then subjected them to combinations of home cleaning procedures like rinsing under tap water and soaking in vinegar or lemon solutions.[6] Soaking for two minutes in lemon juice or vinegar achieved no more bacterial reduction than simply running under the faucet for fifteen seconds. With the exception of broccoli, all the rest of the produce got even cleaner when it was soaked in water for two minutes prior to rinsing in the sink. Of course, brushing or wiping the foods with a paper towel also resulted in a greater reduction in bacterial contamination. In the end, the researchers recommended that consumers rinse fresh produce under cold tap water before consumption, rub or brush the foods clean when possible, and avoid using lemon juice or vinegar, which have no effect.

These studies did not test baking soda because it's widely accepted that this substance has no disinfectant properties. However, sometimes it gets combined with other household treatments. One study into *Salmonella enterica* contamination on spinach leaves confirmed both a chlorine solution's greater disinfectant efficacy when compared to other products and a baking soda solution's total ineffectiveness.[7] Simply rinsing the spinach leaves under the faucet was better than soaking with baking soda.

I know some of you might protest that baking soda manufacturers themselves say their product can be used to clean fruits and vegetables. My response is that, firstly, you need to take everything you read on labels with a pinch of salt, and secondly, these labels state that baking soda can be used to wash (that is, remove dirt from) produce, not disinfect it. Baking soda packaging simply asserts that this substance is safe for use with food, which is true—it's a harmless substance. You could just as easily write "can be used to clean" on table salt or whatever other non-toxic substance you choose. When you apply baking soda to produce, you're subjecting the dirt on your apple to the product's abrasive mechanical action, so if it works, you should really be thanking the sponge you used to rub it around. If you still want to clean your food with baking soda, please do it under running water.

---

6 Agnes Kilonzo-Nthenge, Fur-Chi Chen, and Sandria L Godwin. "Efficacy of home washing methods in controlling surface microbial contamination on fresh produce." *Journal of Food Protection* 69, no. 2 (February 2006): 330-4.

7 Agnes Kilonzo-Nthenge and Siqin Liu. "Antimicrobial efficacy of household sanitizers against artificially inoculated *Salmonella* on ready-to-eat spinach (*Spinacia oleracea*)." *Journal of Consumer Protection and Food Safety* 14, no. 2 (January 2019): 105-12.

TOXOPLASMOSIS

By now, you're aware that we're surrounded by microorganisms. Many live inside us, and others are used to make food and beverages like cheese, yogurt, bread, and beer. Some, like *Salmonella* or *Listeria*, can be dangerous when ingested, so it's important to always practice proper food handling and cooking techniques.

There are some microorganisms, however, that are harmless under normal circumstances. Take *Toxoplasma gondii*, the single-celled parasitic protozoan that causes an infection called toxoplasmosis. The fact that it's a protozoan rather than a bacterium has important consequences, as we'll see later. Protozoa are up to ten times larger than bacteria, and toxoplasmosis is a zoonosis: a disease transmitted from animals to humans.

In the vast majority of cases, these infections are asymptomatic for healthy adults—you wouldn't even know you were infected, or at most, you'd have flu-like symptoms like fever, swollen glands, and muscle pains that would go away on their own. However, the parasite can sometimes cause more serious issues like ocular toxoplasmosis, which is the most frequent cause of uveitis, or inflammation of the eye. And if a pregnant woman gets toxoplasmosis, the infection can cross the placenta and reach the fetus. If the child is born infected (with what we call congenital toxoplasmosis), this can lead to serious problems like miscarriage, cerebral calcification, developmental delays, deafness, and blindness.

It's widely believed that washing fruit and vegetables with baking soda kills toxoplasmosis. Unfortunately, that's not true, since as we've just learned, baking soda does not disinfect. The same misconception exists for chlorine- and hypochlorite-based cleaning products. Although these substances are effective against both harmless and pathogenic bacteria, toxoplasmosis is not a bacterium, so even sodium hypochlorite solutions do nothing against it. However, chlorine-based produce washes can still help pregnant woman avoid infections from bacteria like *Salmonella* and *Listeria*.

Toxoplasmosis and pregnancy

Toxoplasmosis is actually fairly common, although it's now declining thanks to improved sanitation and more hygienic living standards.[8] This is not necessarily a good thing, since if someone has been infected months or years before conceiving, they develop antibodies and are safe from reinfection. If someone contracts the illness for the first time while pregnant, it becomes much more dangerous, and the risks to the unborn baby are far more serious if this occurs in the early months. Cases of congenital toxoplasmosis with severe symptoms typically occur before the twenty-sixth week of pregnancy. As soon as the infection is detected, further tests are carried out, and if

8 Maria Grazia Capretti et al. "Toxoplasmosis in pregnancy in an area with low seroprevalence: is prenatal screening still worthwhile?" *The Pediatric Infectious Disease Journal* 33, no. 1 (January 2014): 5-10.

the illness is confirmed, the mother is treated with antibiotics right away to protect the unborn child. Transmission of the infection through the breast milk of infected mothers has not been documented.

If a pregnant woman gets toxoplasmosis, what is the likelihood she'll pass on the infection to her baby? A study carried out by the University of Catania found that this probability is around 10 percent, which means that nearly one child in five thousand is born with congenital toxoplasmosis.[9] Out of these babies, 1 to 2 percent die or develop neurological deficits, while 4 to 27 percent have permanent visual impairment.[10]

Given the severity of the potential consequences, pregnant women who have not developed immunity to toxoplasmosis should take every step possible to minimize their risk of infection.

Cats as carriers

Asymptomatic infection with toxoplasmosis is common in many species—such as birds, reptiles, and mammals, humans included—but it only fully replicates in felines, in the intestinal tracts of cats in particular. Oocysts (a form of parasite) form in their digestive systems and are then shed in their feces. Infection can take place if you come into contact with these parasites—for example, if you eat contaminated food.

Before you evict Tigger, Felix, and all their fluffy toys, there's something you need to know: Cat feces may be a vehicle for toxoplasmosis oocysts, but pet cats actually play only a secondary role in the disease's transmission—so there's no need to get rid of them.

So, how do cats get toxoplasmosis? From eating infected small prey or uncooked food—for example, an outdoor cat may catch infected birds and mice and become infected in turn. If your cat lives exclusively indoors and you know exactly what it's eating (some brands of cat food even come with guarantees that they're toxoplasmosis-free), the chances of your pet contracting the parasite are very low.

Cats that get toxoplasmosis mostly shed oocysts during their first infection. Many people believe that, after this period is over, cats no longer excrete these parasites—but the truth is, there are reported cases of pets emitting them years later. Oocysts are infectious within two to five days of release and can survive for up to eighteen months in warm, moist environments, while dehydration and cold shorten their lifespans.

Risk factors

A study produced through the cooperation of a large hospital and a major research university in Rome investigated the key risk factors for toxoplasmosis to understand the kinds of things pregnant women should be most conscious of avoiding.[11]

---

9 F. Genovese et al. "Studio epidemiologico sulla toxoplasmosi in gravidanza: benefici dello screening." *Giornale Italiano di Ostetricia e Ginecologia* 33, no. 5 (September-October 2011): 259-63.

10 A. J. C. Cook et al. "Sources of toxoplasma infection in pregnant women: European multicentre case-control study." *BMJ* 321, no. 7254 (July 2000): 142-7.

11 Raffaella Thaller, Federica Tammaro, and Henny Pentimalli. "Fattori di rischio per la toxoplasmosi in gravidanza in una popolazione del centro Italia." *Le Infezioni in Medicina* 4 (2011): 241-7

HOW TOXOPLASMOSIS IS TRANSMITTED

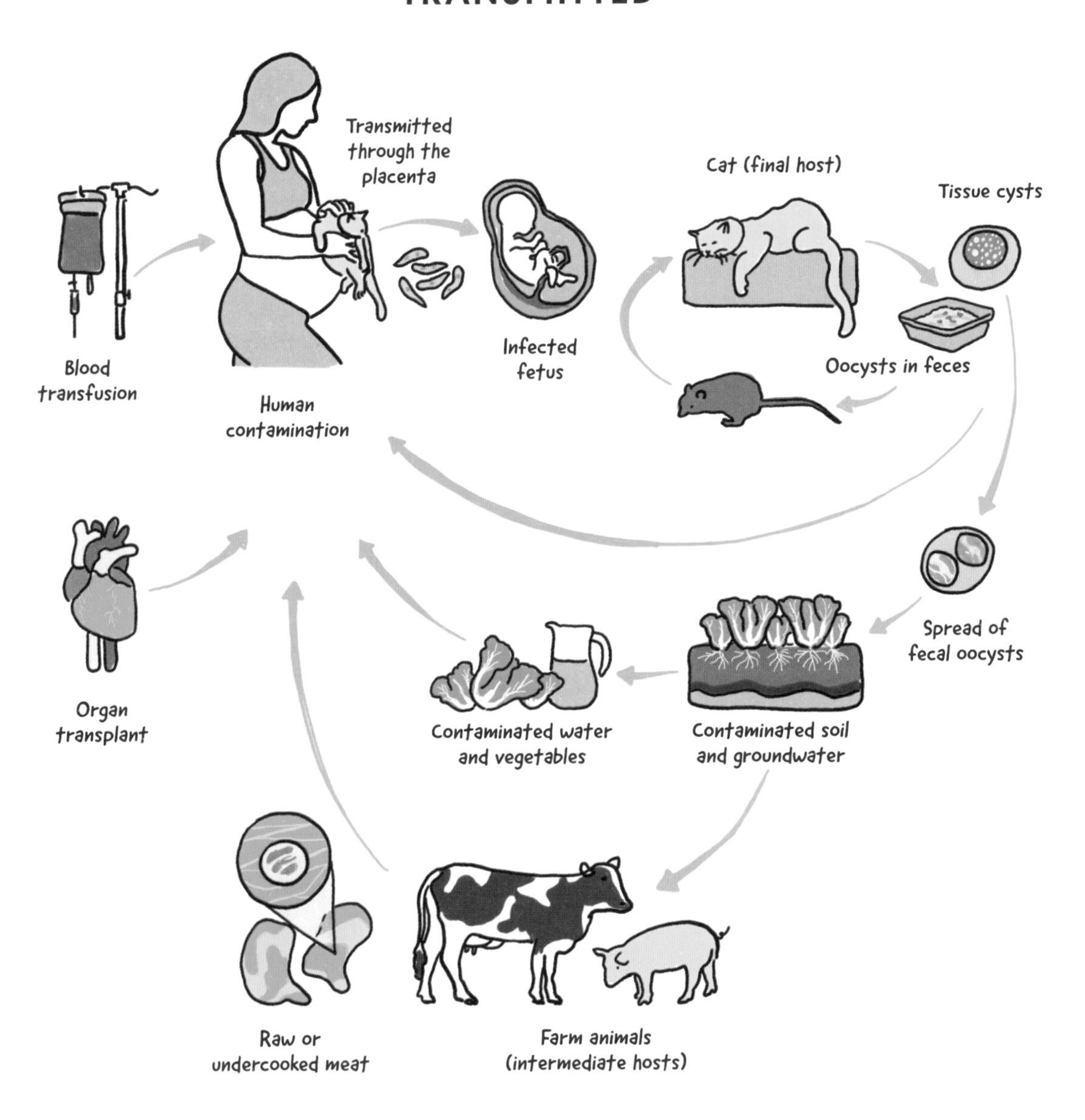

Sausages, salami, and other cured meats are the most likely source of infection, especially those prepared at home or by small manufacturers, although mass-produced varieties also present potential hazards. Handcrafted sausages are often not cured for as long as brand-name meats and may contain less salt. They also may be made from intestines, tongue, and offal, which contain more cysts than the muscles of infected animals. Also, animals on small farms often live outdoors and are more likely to become infected.

Additionally, eating raw or undercooked meat can lead to infection. If you're pregnant, you should avoid sushi, rare steak, and foods like steak tartare, since these can contain not only toxoplasmosis, but also dangerous bacteria. Also, remember never to taste uncooked dishes like meatloaf or ravioli filling to check if they need more seasoning. Finally, raw dough from baked goods can contain other harmful non-toxoplasmosis bacteria, so you should avoid sampling cakes or cookies before putting them in the oven.

Although toxoplasmosis can infect the muscles of any animal we eat, it dies if meat is cooked for five minutes at a temperature above 145°F (63°C). Freezing meats also reduces toxoplasmosis's ability to cause infection, but there's no evidence that it actually kills the parasite's various forms.[12] The same applies to curing cold cuts: There's no clear proof that a long curing process eliminates the risk. It should also be noted that freezing does not kill bacteria—instead, they become inactive in the freezer, then start multiplying again once thawed.

Gardening and growing plants or living in rural areas can also potentially lead to a toxoplasmosis infection. Oocysts can happily reside in soil for up to fifteen months, so you could become infected by accidentally touching contaminated dirt or sand before putting your hand on your mouth. As we've learned, living with cats does carry risk, but researchers are inclined to think that this is connected to felines' contact with the soil—so any type of outdoor pet, especially dogs, is a potential source of infection.

Eating raw fruit and vegetables is a risk factor, but much less so than the previous items I've listed, because thoroughly washing produce or eating only industrially packaged products reduces the chance of contagion. Rinsing produce under running water is always useful because it washes away soil that could contain oocysts, while soaking is not recommended as it just spreads the dirt everywhere in the solution. Carefully cleaning fruits and vegetables can also stave off infection from any type of harmful bacteria—but I want to remind you that neither chlorine-based washes nor baking soda can kill toxoplasmosis. Of course, not all produce presents the same risk: If you grow strawberries in your garden, they're more likely to be exposed to cat feces than apples you've picked from your tree.

12 A. Mantovani, B. Cacciapuoti, and A. Ioppolo. "Prevenzione della toxoplasmosi di origine animale." *Annali Dell'istituto Superiore di Sanità* 20, no. 4 (1984): 313-6.

PESTICIDES

The chemical or biological substances known as pesticides are an integral part of modern agriculture that's used in the cultivation of a whole host of foods, from grains and pulses to fruit and vegetables. They protect crops from pests like fungi, bacteria, rodents, and insects. In today's industry, it's impossible for traditional or organic farmers to do without pesticides, so it's legitimate to be concerned about whether there's potentially harmful residue on the products we purchase and eat.[13] As a remedy, we're often advised to soak fresh produce in water and baking soda—and in this particular instance, this could be good advice.

Newspaper and magazine articles that make this recommendation usually cite a study in which researchers applied two pesticides[14] to the surface of an apple, then tried to remove them the next day by either rinsing the fruit under the faucet for two minutes or soaking it in water and baking soda at a concentration of 10 grams per liter.[15] Washing apples in the sink reduced fungicide and insecticide residues by around a third and a quarter. Soaking in baking soda was much more effective: The apples were 100 percent residue-free after sitting for twelve and fifteen minutes respectively in water and baking soda and then getting a short rinse. However, small amounts of pesticide had penetrated the peel—20 percent of the fungicide and 4.4 percent of the insecticide—and were not removed. To get rid of all traces of pesticide, including from the apple's flesh, you would have to peel it before eating. The downside is that you would then lose out on the many beneficial substances, such as polyphenols, that this outer layer contains.

The results of this study are interesting, but we have to be careful about generalizing them to other pesticides and foods, because each combination has its own idiosyncrasies. The type of produce is key: A knobby head of broccoli is a much more complex challenge than a smooth apple. On top of this, the study's experimental conditions were in no way representative of the real world. The researchers washed the apples the day after adding the pesticides, but in reality, there are regulations about minimum times between applying pesticides and harvesting crops. Rules like these ensure consumers aren't exposed to potentially harmful levels of residues. Additionally, as the researchers explained, the two pesticides they used spontaneously degrade when exposed to the alkalinity of baking soda, making them easier to remove with rinsing and soaking. Other materials may break down more quickly at an acidic pH or in the presence of different substances, like sodium hypochlorite (which was found to be ineffective in this study).

---

13 Regulated organic farming practices allow a limited number of pesticides to protect crops.

14 Thiabendazole, a fungicide, and phosmet, an insecticide.

15 Tianxi Yang et al. "Effectiveness of Commercial and Homemade Washing Agents in Removing Pesticide Residues on and in Apples." *Journal of Agricultural and Food Chemistry* 65, no. 44 (October 2017): 9744–52.

Water might actually be the best substance for removing residues, as it's an excellent solvent. There was a similar study in which two different pesticides[16] were applied to Chinese cabbage and broccoli.[17] In this case, water was the most effective at reducing residue levels in three out of four combinations, while soaking in baking soda came out on top only for Chinese cabbage that had been sprayed with one of the two substances.

But closer reading of two recent articles that reviewed research into removing pesticides from fresh produce reveals that water is not actually the best solution, as it often fails to get rid of anything at all.[18] This is primarily because many plants are often covered with a water-repellent, lipophilic surface film, and since pesticides are also lipophilic, they prove very difficult to dislodge. Based on many hundreds of tests carried out with not only baking soda but also many other substances like salt, sodium hypochlorite, hydrogen peroxide, detergents, acetic acid, and lye, the conclusion is that there's no single treatment that can always be relied on to banish chemical residues. The only 100-percent effective method is peeling produce. However, while we can easily peel apples, we

16 Chlorpyrifos and cypermethrin.

17 Miao-Fan Chen et al. "Insecticide residues in head lettuce, cabbage, Chinese cabbage, and broccoli grown in fields." *Journal of Agricultural and Food Chemistry* 62, no. 16 (March 2014): 3644–8.

18 Stephen W. C. Chung. "How effective are common household preparations on removing pesticide residues from fruit and vegetables? A review." *Journal of the Science of Food and Agriculture* 98, no. 8 (June 2018): 2857–70.
Tanmayee Bhilwadikar et al. "Decontamination of Microorganisms and Pesticides from Fresh Fruits and Vegetables: A Comprehensive Review from Common Household Processes to Modern Techniques." *Comprehensive Reviews in Food Science and Food Safety* 18, no. 4 (July 2019): 1003–38.

certainly can't peel spinach. And what a shame it would be to throw away the many healthy substances found in the outer casing.

So, what should we do? Let's take a step back and ask ourselves this question first: Why do we want to remove pesticide residues? Just because they're there? At this point, we need to go back to the difference between risk and hazard from chapter 1. Worrying about pesticides on your food is like trying to kill a shark without first knowing whether you're watching it from a clifftop or swimming in the water right next to it. Undoubtedly, some pesticides are health hazards. We've all come across alarmist news articles that list the many different pesticides found on this or that fruit or vegetable. However, any action we take shouldn't be driven by instinctive fear, but rather a rational understanding of the actual situation—so if you're worried, the right question to research is, "What risk do these residues pose to me?" It's important to remember that legislation sets limits on pesticide residues to ensure they do not present an unacceptable risk to consumers.

12

Household Surfaces

Our homes have lots of hard surfaces to clean—from the stovetop and the kitchen floor to rugs and carpets, from tables and cabinets to windows and mirrors. This far into the book, I'm sure you know what I'm about to say: The detergent you choose for a particular surface must have the right combination of ingredients (such as surfactants, solvents, water, abrasives, and acids or bases) for the kind of dirt and type of material you want to remove it from.

Unlike laundry detergent and dish soap, surface-cleaning products must work at room temperature, usually with very short contact times—and, as always, without damaging whatever it is we want to clean. On the grocery store shelves, you'll find both all-purpose cleaners and more specific substances suitable for getting one type of dirt off one particular area. Surfactant concentration varies from less than 1 percent to more than 30 percent, with pH ranging from highly acidic to highly basic, and cleaners come in many forms: liquid, foam, powder, granule, spray, cream, and more.

There's no point in going through them one by one, since many work in the same way as and feature ingredients that are similar to the laundry products, dishwasher detergents, and stain removers we've already seen. Instead, I'll just touch on guidelines for a few household surfaces.

FLOORS

Floor-cleaning products have been with us for more than a century now, and the earliest versions were pretty basic: no more than a simple soap and ammonia solution. The ammonia dissolved the grease, and the surfactants in the soap carried it away with all the other grime. A few modern products still contain ammonia—you can easily identify it by its unpleasant smell—but as they've evolved, many subcategories and specializations have been added. Beyond advances in ingredients, there have also been improvements to the cloths and mops they're applied with. Special textiles like microfiber are becoming increasingly popular—their main benefit is that their increased surface area helps them capture far more dirt than ordinary cotton.

The range of floor materials is endless (including wood, concrete, PVC, marble, linoleum, and ceramic), so you need to be sure that the product you're using won't damage the surface. For example, acidic detergent isn't recommended for marble or limestone. Generally speaking, floor cleaners have neutral or

DID YOU KNOW?

Bathroom cleaners that are too acidic will eventually corrode tile grout. That's why these types are not often recommended—vinegar included! If acidity is absolutely necessary for the job at hand, then it's much better to use something less problematic, like phosphoric acid.

mildly basic pH values. (Specific toilet cleaners are the only highly acidic products designed for surfaces.)

Before you apply a product to your floor, it's always best to vacuum up any dust or stray solids, especially since otherwise, the latter can end up scratching the surface. Back in the day, floors also had to be rinsed after washing to prevent ingredients like surfactants from leaving behind residue after the floor dried. Nowadays, most products don't need to be rinsed—formulations have been tweaked, solutions have been diluted, and organic solvents, which evaporate without a trace, have been added.

Today's solvents are highly effective at dissolving dirt, transferring it to a cloth or mop, and then evaporating to leave behind a pleasant smell. Common varieties are limonene, which comes from the oils produced by processing oranges into juice, and pine essential oil, which is extracted from natural sources and is therefore often preferred by consumers who are wary of "chemicals." Sometimes, these solvents are added in doses too small to serve any cleaning purpose beyond simply smelling nice. In this case, the dirt-removal heavy lifting is left to other categories of ingredients with solvent properties, such as glycol ethers.

It's also becoming increasingly popular to include value-added ingredients that offer more than just cleaning power. For example, detergents for floor materials that are easily damaged (like parquet) feature substances that hide scratches by smoothing them over or filling them in with a polymer or resin[1] that becomes shiny once it dries.

Mops

Many of us probably have an old-fashioned mop at home: a long pole with a bundle of fabric, yarn, sponge, or other absorbent material attached that we dip into water and detergent. The wet fabric spreads the solution across the floor, and the rubbing action detaches the dirt, which gets trapped in the mop's fringes. Then, you wring out the mop into the bucket, dip the fringes back in to get more water and detergent, and keep going until the floor is clean, changing out the liquid when it gets too dirty.

Professional cleaners recommend using two or even three buckets to get the environment as clean and free of residue as possible. With two buckets, you rinse the mop in the container filled with clean water only, squeeze it out, then dunk it into the bucket of cleaning solution again. In the three-bucket procedure, the mop gets rinsed in one bucket of clean water, then wrung out in another. This keeps the washing water relatively clean, preventing you from spreading around the dirt you've already picked up.

1 Such as acrylate polymers, polyvinyl alcohols, polysaccharides, or PEG (polyethylene glycol).

GLASS AND MIRRORS

I don't know if this ever happens to you, but if I'm looking out the window and spot a mark on the glass, I have an irresistible urge to get up and clean it. Sometimes, it turns out to be a dead insect, and other times, a fingerprint—probably my own from the day before. I realize this may seem a bit weird, but I feel a glow of satisfaction when I restore the window to its original luster and transparency. The same sense of accomplishment comes from getting rid of those pesky toothpaste marks on the bathroom mirror or the annoying white streaks on the shower door.

When cleaning the countless glass and otherwise transparent surfaces in our homes, we need to go back to a concept I've emphasized throughout this book: First, think about the type of dirt. The grime on the insides of my kitchen windows is probably grease. I regularly sauté vegetables in oil, meaning that often, the air becomes filled with tiny drops of oil that escape being whisked away by the range hood and instead end up on surfaces around the kitchen. Their removal requires a detergent—even just a drop of dish soap in some water. The dirt on the outsides of my windows, on the other hand, is more likely to be made up of dust and other fine airborne particles, maybe with some squashed dead insects mixed in. Water and alcohol are my go-to products for that job. If your kids have doodled on the window glass with markers, you can either preserve their masterpieces . . . or wash them away with an organic solvent like ethyl alcohol, which is more effective than a detergent for this task.

When I served in the military, I often had to clean our dorm's long bathroom mirrors. My sergeant showed me what to do: First, wipe the glass with a sponge dipped in soap and water until it's wet, but not dripping.

Second, dry it with a fine cloth, and third, remove any streaks by wiping the surface again with a crumpled sheet of newspaper. The mirrors would be spotless, and the sergeant was always quite pleased. I never really stopped to wonder why the newspaper worked so well, but in hindsight, it must have been a combination of the paper's abrasive effect and the solvent in the ink, which was oil-based back then. (Unfortunately, I don't know if the ink in today's newspapers would still work.)

The white streaks on your glass shower door are probably limescale left behind after the shower dries. In this case, as we learned in chapter 3, acidic cleaners are ideal—a solution of water and either 5-percent vinegar or 3-percent citric acid (which is my preference, since it doesn't smell) will do the trick.

What can glass cleaners do?

When it comes to cleaning these types of surfaces, the products you're looking for are most often sold as glass cleaners. However, their very basic formulations mean they can actually be used in a variety of contexts beyond just windows, mirrors, or tiles. These cleaners are great for light dirt that accumulates around your kitchen and bathroom, and they're also suitable for hard plastic surfaces like furniture, kitchen appliances, and doors. It's always best to do a spot test first, though, as their solvents can ruin some plastics. These products also don't work on more stubborn dirt (so that's a no to using them on dirty dishes) or aluminum (which they can damage).

Most glass cleaners are packaged as sprays, and they were actually the first detergents ever sold in the spray bottles we've now become so accustomed to. This packaging was an instant hit thanks to how effectively and accurately it can apply cleaning liquids to areas that would otherwise be tricky to reach with a sponge and a bucket of water, like tall vertical surfaces. Sprays convert liquids into a fine mist made up of hundreds of miniscule droplets that cling to a surface instead of dribbling downward. If you find that the product runs more than you'd like, you're probably standing too close to your target or using too much.

If you have glass or a hard surface in your home that's become covered in thick layers of grease—maybe the range hood, for example—I recommend detaching it if possible, then washing it in the sink with a sponge and dish soap. When it's dry, remove any streaks with a glass cleaner. In some cases, though, glass cleaners are completely useless: For instance, they do nothing for limescale buildup on your shower door, so instead you'll need to try one of the acidic solutions I described earlier.

Some products contain anti-fog agents that prevent surfaces like mirrors from fogging up. Manufacturers have two completely different ways of doing this: Some add polymers that spread into a thin film across the glass, creating a hydrophilic surface—when water touches the mirror, it remains a continuous film rather than breaking up into droplets that create the misty appearance. Others do the opposite, using hydrophobic substances like silicon that repel water droplets, preventing them from depositing on the mirror in the first place.

PRODUCT SPOTLIGHT: **GLASS CLEANER**

Manufacturers don't know exactly what kinds of dirt their glass cleaners will have to remove, but they can assume that the surface will be only lightly soiled, so the detergent doesn't need to contain any highly aggressive cleaning agents. The one all-important property for products in this category is that they absolutely must not leave behind streaks, spots, or smears. This means they can't contain salts or substances that leave solid residues, and the overall product must fully evaporate and easily get absorbed by the cloth used to wipe away the dirt after spraying.

So, it should come as no surprise that the main ingredient is water, often making up 90 percent of the formulation—glass cleaners are some of the most highly diluted cleaning products on the market. But don't get me wrong: Manufacturers aren't trying to con you by pushing water under a fancy name (as some Facebook cleaning pages may tell you). On its own, water is an excellent cleaner, and it's also a solvent that's useful for diluting substances required to clean types of dirt that it can't tackle alone. And water evaporates fully, leaving absolutely no trace.

After water, the first ingredient in many cleaning products is a surfactant . . . but not in glass cleaners. Their second ingredient is an organic solvent, which can be an alcohol like ethanol or isopropyl—the latter is used to make disposable glass and screen wipes.[2] Alcohols have three key properties that make them ideal for cleaning glass: They dissolve greasy grime and other substances like ink, they evaporate without a trace, and they're completely soluble in water.

For even more effective cleaning, a source of alkalinity can be useful. The preferred pH-raising substance in the glass-cleaning category is ammonia, a powerful cleaner that's also a volatile chemical (meaning it leaves nothing of itself behind). Unfortunately, many people (including me) think it smells really bad. This malodorous substance is a good solvent, so it removes any greasy dirt. When I checked the online ingredient lists of cleaning products containing ammonia, I found that they always have less alcohol than ammonia-free cleaners.

The final key components in glass cleaners are surfactants—generally nonionic, but there are some anionic varieties in concentrations below 1 percent. They're also used to lift small traces of oily dirt, often in the form of fingerprints. The quantity of surfactants needs to be just enough to enable the water to wet the surface well and remove the few fats present without foaming.

Like many other products, glass cleaners also often include preservatives, fragrances, and colorants. Colorants are important because they not only make it easier and safer to apply the product by helping you see where it's been sprayed, but also prevent the liquid

2 Sometimes glass cleaners contain less common solvents such as glycol ethers or more complex alcohols like 1-methoxy-2-propanol or methoxyisopropanol.

from being mistaken for water and accidentally consumed. (You may think this could never happen, but the number of people admitted to the hospital every year with poisoning or intoxication from unintentional ingestion of a cleaning product is shockingly high.)

Can I make it at home?

As I've said before, I'm not a fan of homemade cleaning products. It's not as easy as it might seem to replicate the specific functions chemists have spent weeks, months, or even years developing in a laboratory, and it can even be dangerous depending on the substances involved. But, if you're still interested in giving it a try, you can actually mix up a glass-cleaning product quite easily if you don't add fragrances or colorants and are fine with your mixture lasting only for a little while without a preservative.

You can formulate your window cleaner in an empty spray bottle. Always be sure to label the cleaning products you make, and to avoid unpleasant accidents, never use bottles that you also drink from.

Commercial glass cleaners are often 5 to 10 percent alcohol. For your homemade substance, try ethyl alcohol. The denatured pink variety can leave a reddish film on glass due to the dye it includes, so it's better to use the food-grade clear version (which is more expensive—but DIY products aren't usually as cheap as you'd expect). Put 5 to 10 grams of alcohol in your spray bottle, then top up it with water until it weighs 100 grams total.

What kind of water should you use? Since glass cleaners must not smear, I don't recommend tap water, especially in areas with hard water where limescale is more common. Deionized water, which is available at any supermarket, is much better. The very best option is a small bottle of natural spring water with a low fixed residue.

You also need a surfactant—for this, you need look no further than ordinary dish soap. A few drops diluted in your spray bottle are all you really need, since anything more would generate too many bubbles and leave a film on the surface after the substance dries. If there are any streaks or smears after you clean, the culprit is very likely this detergent, which definitely wasn't formulated to clean glass. Next time, you can try a different dish soap . . . or spend a couple of dollars on a commercial product specifically designed for your purpose by trained chemists.

HOMEMADE GLASS CLEANER

CARPETS AND RUGS

The house where I grew up had a lovely carpet—its shade of orange was warm, but not too bright, and I especially loved how soft it felt under my feet as I padded around (almost always) barefoot. I was a child in the 1970s, when carpet was all the rage. The enthusiasm faded over time, though, mainly because of how difficult it was to keep clean. Many houses later, I no longer have carpet, just wooden floors and tiles that I can vacuum and mop without having to invest in any extra appliances, foam cleaners, or other special technologies.

The difficulties my mother Iride faced back in the 70s are not dissimilar to those experienced today by anyone who has a carpet or rug at home. Smaller rugs can be popped into the washing machine if the fabric allows (but you should vacuum away any solid dirt first), whereas bigger ones or those with non-machine-washable backings require different techniques, from specialized foams and shampoos to a simple scrubbing brush and a dab of detergent.

Dirt on a rug or carpet can be located at the tips or bases of the threads, or it can even penetrate the backing. When you wash a soiled piece of clothing, first the dirt gets detached from the wet fabric, then the water evaporates and the garment dries. This process is more or less impossible to achieve with a carpet, and it's rarely a good idea to get the base wet. Instead, some products lift dirt from the base and fibers so that the grime can be easily picked up by a vacuum cleaner once the liquid has evaporated.

Washing with very bubbly water can sometimes help dislodge dirt, which can then be fully removed with the help of other means, like a carpet cleaner. This might be the only case in which a vinegar and baking soda concoction makes any sense, precisely because of the carbon dioxide bubbles it yields. You can sprinkle baking soda onto the dirty area, then pour on some vinegar. (You know by now that I prefer odorless citric acid, but if you want to use vinegar, just make sure you open the windows to ventilate the room afterward.) However, if the dirt you're dealing with is oily or inky, then none of what I've just said can help, and you'll need to invest in a specialized cleaning product.

DID YOU KNOW?

Many cleaning products smell like lemons because manufacturers are capitalizing on the widespread belief that lemon juice and oil have detergent properties.

FURNITURE

Furniture cleaning products are often formulated to both clean and polish, especially if they're intended for wood. The layer of dirt on these surfaces is usually very fine, often nothing more than dust. When these cleaning products are applied, they leave behind a thin coat of oil, wax, resin, or silicon.

STOVETOPS AND OVENS

The surfaces that are probably most at risk of getting coated with a buildup of food are the stovetop and the inside of the oven. Stove burners get caked with all kinds of dirt very quickly—which is why from the minute this technology was introduced, any products designed to clean them have always had a powerful scrubbing capability. Initially, powders that had to be dissolved in water were the most popular choice, generally featuring a basic detergent and an abrasive (usually silica).

These were followed by cream cleaners, which won over consumers. Like their predecessors, they also contained a solid abrasive suspended in a liquid, but as surfaces became more delicate over the years, less-harsh materials were required. Nowadays, calcium carbonate or other substances that are much softer than silica or quartz (like oxides or aluminum silicates) are used. Interestingly, consumers tend to see cream products as more delicate than powder varieties, even when they include equivalent ingredients.

The type of abrasive is just as important as the shape and size of the solid granules dispersed in the cream. This category of product is often employed not just in the kitchen and on ceramic bathroom fixtures, but also on more delicate surfaces like methacrylate shower walls. Be careful while doing this, though: Try out your product on a less visible corner of the shower first.

You probably wipe down your stovetop fairly regularly, but let's be honest: How often do you clean your oven? That's why the walls inside are most likely the dirtiest area of your home. The main source of the crusty residue on the inside walls and glass door is baked-on fat. When oils or fats are heated to a high enough temperature, they transform through a process called polymerization and become almost impossible to remove. I'm sure you've seen those slick, brownish-yellow hardened marks that no amount of scrubbing with a detergent can budge.

Since they have to deal with such stubborn grime, oven cleaners are nearly the most aggressive products of all (with drain cleaners, discussed on page 17, coming in first place). They often contain lye or other powerful chemicals, and they're typically sprayed on or applied as a foam. Be careful when using oven-cleaning products: Always wear safety gloves and goggles, and read the instructions thoroughly. I'd also be wary of self-made abrasive powder solutions, which

often feature—you guessed it—baking soda. This substance shouldn't cause issues if you're cleaning baking trays, but oven walls usually have a special coating that can be damaged by an abrasive, so make sure you read your oven's instructions before you start cleaning.

DRAINS

Although it's not technically a household surface, any respectable book about cleaning can't avoid mentioning the one thing we all have to deal with at some point in our lives: a blocked drain! It may allow water to trickle in very slowly, or it might be clogged up completely. You know the procedure by now: Analyze first to work out what's causing the block, then decide on the most sensible scientific solution. And before we go any further, let's get one thing straight—no, vinegar and baking soda definitely won't work.

If you have a blocked drain in your kitchen or bathroom, the first thing you need to do is have a look underneath the sink. Unscrew the U-bend (which is usually made of plastic), look for any clumps of food or other nasty substances and empty them out, then put the pipe back together again. In most cases, this is enough to solve the problem. You could call a plumber, but it's really quite simple to do by yourself.

Real issues occur when the blockage is somewhere you can't reach, maybe at a point in the pipe that's covered by the wall. This is when you need to track down the plumber's number—although before you dial it, there is one more thing you can do. For the following techniques to work, some water must still be able to flow down the pipes. If nothing at all is getting through, as a last resort, you can get out your trusty plunger and use it to force air in and out of the drain, hopefully dislodging the cause of the blockage.

There are three types of residue that can slow a fast-draining sink to a gradual trickle. The first is limescale. As we've discussed several times already, if you live in an area where the water is rich in calcium and magnesium, the resulting carbonates can build up on the insides of your pipes, gradually slowing down the fluid. As the diameter of the pipe is reduced, small solids can get trapped more easily. This is a slow process that doesn't happen overnight, so it can't explain a sudden blockage—for several years leading up to the issue, you would have noticed a gradual reduction in flow.

The second thing that can impede or completely stop water from draining is protein, whether this comes from food residue or other substances like animal or human hair. Most proteins don't dissolve, even in boiling water—if you try pouring some over a few strands of hair, you'll notice that nothing happens. If the water in your area is mineral-rich and some scraps of food get into the drain, they'll combine into a paste containing proteins solidified in calcium carbonate.

A third issue that can compound with all of the above to block your sink is fat, especially solid fat. When I made spaghetti carbonara for lunch yesterday, I fried the pork and drained the softened fat into a glass. By afternoon, it had completely returned to its original solidified state. If I'd poured it down the sink in its liquid form (please, don't ever do this!), it would have gradually hardened and potentially plugged up the pipe.

So, the main causes of drain blockages are limescale, protein, and fat, and no single chemical substance on the planet can eliminate all three at once. I'd especially like to draw your attention to the fact that neither salt nor baking soda can help with these in any way, since both are completely inert with regard to all three substances.

If your blockage comes from one of these sources and water is still trickling past, however slowly, then there's still hope that we can do something about it. But first, we have to work out which material is impeding the flow. Let's start with limescale, which binds

HOW TO DEAL WITH A BLOCKED DRAIN

IF THE RESIDUE IS . . .	THEN TRY UNCLOGGING WITH . . .
Limescale	Warm citric acid or vinegar
Protein (food residue or hair)	Lye
Fat	Hot water and dish soap

all the others together. By now, you know that to get rid of an alkaline substance like limescale, we need an acid. Vinegar can work, and citric acid is even better because it's stronger, doesn't corrode rubber seals, and is gentler on chrome finishes. Whichever substance you choose, heating it up before you pour it down the drain makes it more effective at dissolving limescale. Warm acids are also useful against alkaline residues from soap or detergent.

Let's turn our attention to clumps of protein now. Lye (that is, sodium hydroxide) is our weapon of choice here—it's a hazardous substance that's the main (and sometimes the only) ingredient in products that are made for unblocking drains. When it's sold by itself, lye comes in gel, flake, or microbead forms. Before using this product, check if any of your pipes are aluminum, since it damages this substance (but doesn't affect plastic). Also, follow the instructions to the letter and always wear safety glasses. The pH of the lye you pour down the drain might be higher than 13, which explains why baking soda, which is only mildly basic, is useless in the same situation. As well as attacking proteins, lye turns fats into soap, rendering them soluble.

If you're dealing with this third potential culprit, solid fat can also be cleared by pouring a couple of pots of water heated to at least 140°F (60°C) down the drain. The warm vinegar you sent down earlier may even have started softening it already. With this technique, there's a risk of just shifting the blockage further along the pipe—so ideally, you should add some soap or detergent (like dish soap) to the water. This will emulsify the fat and ensure it's carried away by the hot water before it solidifies again.

By now, if the blockage was caused by limescale, protein, or fat, it should be cleared away. But if all else fails, you can buy commercial drain cleaners containing sulfuric acid: an acid so strong that it makes light work of all organic matter, protein and fat included. This substance is dangerous, though, so I don't recommend it—it's too risky. In my opinion, the time has come to call in the plumber.

Now do you understand why that YouTube video I mentioned in the introduction about unblocking your sink with salt, vinegar, and baking soda offered such useless advice? People who followed its instructions jammed up any unclogged space left in their drains, which could have instead been filled with a substance capable of clearing the obstruction. If this trick appeared to work for anyone, it was probably thanks to large quantities of warm vinegar—after it made short shrift of the baking soda, there was still enough left to remove limescale and melt fat. There definitely couldn't have been any protein in the YouTuber's drain.

As always, analyzing the problem is the best way to find an effective solution—or at least give it your best shot. It's a plumber's job to know all this, so I haven't invented anything new. All it takes is using chemistry properly . . . and letting go of the belief that everything can be fixed with a mixture of vinegar, baking soda, salt, and lemon.

ACKNOWLEDGMENTS

Products featured in the Product Spotlight sections were selected without the knowledge of the respective manufacturers. I chose them either because they're particularly well-known, even iconic, or because they have especially simple or easy-to-interpret formulations. Sometimes, I picked them just because they were easier to research than their rivals. The explanations provided for each ingredient are my own, including any errors in the reasoning for their inclusion—it's not always easy to retrospectively understand why a particular substance might have been added. In any case, the inclusion of any product in this book is by no means an endorsement or advertisement.

Thanks to Beatrice Mautino and Alessandro Mustazzolu for reading and editing several chapters of the original Italian edition of this book. Any mistakes that may remain are my own. I would also like to thank all the people who have interacted with me on social media (YouTube, Instagram, TikTok, Facebook, and X) for the incredible number of tips and interesting facts they have shared with me, not to mention the warmth they have shown me at a very difficult time in my life. If I have achieved any clarity in this book, it's also thanks to them.

SELECTED BIBLIOGRAPHY

Cutler, W. Gale and Erik Kissa, eds. *Detergency: Theory and Technology*. New York: Marcel Dekker, Inc., 1987.

Lai, Kuo-Yann, ed. *Liquid Detergents*. 2nd ed. London: Taylor & Francis, 2006.

Smulders, Eduard. *Laundry Detergents*. Weinheim: Wiley-VCH, 2002.

Spitz, Luis, ed. *Soap Manufacturing Technology*. 2nd ed. Cambridge, MA: Academic Press and AOCS Press, 2016.

Zoller, Uri, ed. *Handbook of Detergents*. Vols. A–F. Boca Raton: CRC Press, 1999–2008.

INDEX

NOTE: Page references in *italics* refer to illustrations. Page references followed by "n" refer to footnotes.

T

ABOUT THE AUTHOR

DARIO BRESSANINI, PhD, is a chemist, science communicator, and YouTuber. He is a professor at the Department of Science and High Technology of the University of Insubria in Como, Italy, where he teaches and conducts research. In his home country, Bressanini has a large social media following and has published several bestselling books about the science of everyday things. He also writes monthly articles about chemistry in the kitchen for the Italian edition of *Scientific American* (*Le Scienze*).

dariobressanini | dario.bressanini

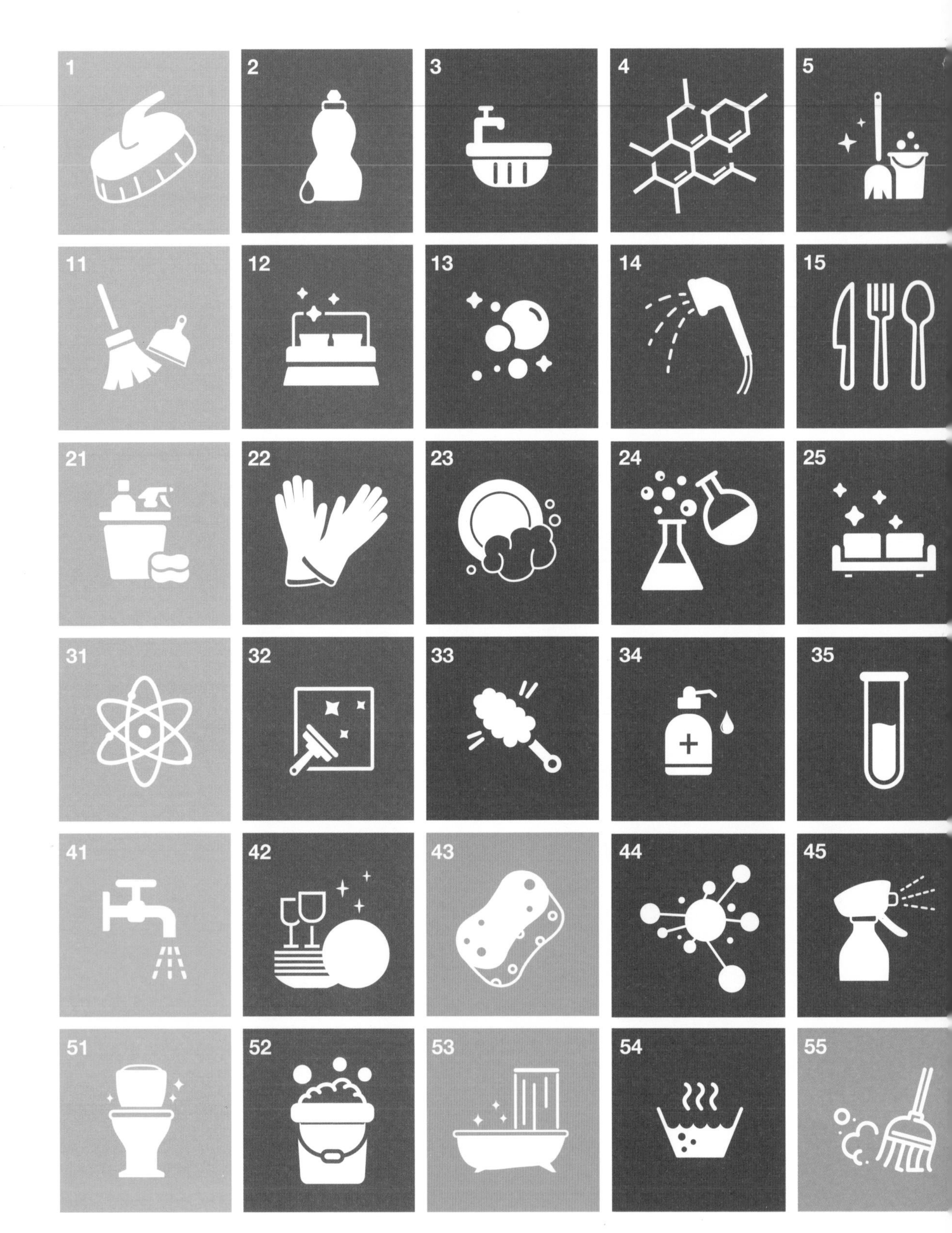
1
2
3
4
5
11
12
13
14
15
21
22
23
24
25
31
32
33
34
35
41
42
43
44
45
51
52
53
54
55